GABOR AND WAVELET FRAMES

LECTURE NOTES SERIES
Institute for Mathematical Sciences, National University of Singapore

Series Editors: Louis H. Y. Chen and Ka Hin Leung
Institute for Mathematical Sciences
National University of Singapore

ISSN: 1793-0758

Published

Vol. 1 Coding Theory and Cryptology
edited by Harald Niederreiter

Vol. 2 Representations of Real and *p*-Adic Groups
edited by Eng-Chye Tan & Chen-Bo Zhu

Vol. 3 Selected Topics in Post-Genome Knowledge Discovery
edited by Limsoon Wong & Louxin Zhang

Vol. 4 An Introduction to Stein's Method
edited by A. D. Barbour & Louis H. Y. Chen

Vol. 5 Stein's Method and Applications
edited by A. D. Barbour & Louis H. Y. Chen

Vol. 6 Computational Methods in Large Scale Simulation
edited by K.-Y. Lam & H.-P. Lee

Vol. 7 Markov Chain Monte Carlo: Innovations and Applications
edited by W. S. Kendall, F. Liang & J.-S. Wang

Vol. 8 Transition and Turbulence Control
edited by Mohamed Gad-el-Hak & Her Mann Tsai

Vol. 9 Dynamics in Models of Coarsening, Coagulation, Condensation and Quantization
edited by Weizhu Bao & Jian-Guo Liu

Vol. 10 Gabor and Wavelet Frames
edited by Say Song Goh, Amos Ron & Zuowei Shen

Lecture Notes Series, Institute for Mathematical Sciences,
National University of Singapore Vol. 10

GABOR AND WAVELET FRAMES

Editors

Say Song Goh
National University of Singapore, Singapore

Amos Ron
University of Wisconsin-Madison, USA

Zuowei Shen
National University of Singapore, Singapore

NEW JERSEY • LONDON • SINGAPORE • BEIJING • SHANGHAI • HONG KONG • TAIPEI • CHENNAI

Published by

World Scientific Publishing Co. Pte. Ltd.
5 Toh Tuck Link, Singapore 596224
USA office: 27 Warren Street, Suite 401-402, Hackensack, NJ 07601
UK office: 57 Shelton Street, Covent Garden, London WC2H 9HE

Library of Congress Cataloging-in-Publication Data
Gabor and wavelet frames / editors Say Song Goh, Amos Ron, Zuowei Shen.
p. cm. -- (Lecture notes series, Institute for Mathematical Sciences, National University of Singapore ; v. 10)
Includes bibliographical references.
ISBN-13: 978-981-270-907-3 (hardcover : alk. paper)
ISBN-10: 981-270-907-X (hardcover : alk. paper)
1. Gabor transforms. 2. Wavelets (Mathematics) I. Goh, Say Song. II. Ron, Amos. III. Shen, Zuowei.

QC20.7.F67.G33 2007
515'.723--dc22

2007019905

British Library Cataloguing-in-Publication Data
A catalogue record for this book is available from the British Library.

Printed in Singapore.

CONTENTS

FOREWORD

The Institute for Mathematical Sciences at the National University of Singapore was established on 1 July 2000. Its mission is to foster mathematical research, particularly multidisciplinary research that links mathematics to other disciplines, to nurture the growth of mathematical expertise among research scientists, to train talent for research in the mathematical sciences, and to provide a platform for interaction and collaboration between local and foreign mathematical scientists, in support of national development.

The Institute organizes thematic programs which last from one month to six months. The theme or themes of a program will generally be of a multidisciplinary nature, chosen from areas at the forefront of current research in the mathematical sciences and their applications, and in accordance with the scientific interests and technological needs in Singapore.

Generally, for each program there will be tutorial lectures on background material followed by workshops at the research level. Notes on these lectures are usually made available to the participants for their immediate benefit during the program. The main objective of the Institute's Lecture Notes Series is to bring these lectures to a wider audience. Occasionally, the Series may also include the proceedings of workshops and expository lectures organized by the Institute.

The World Scientific Publishing Company has kindly agreed to publish the Lecture Notes Series. This Volume, "Gabor and Wavelet Frames", is the tenth of this Series. We hope that through the regular publication of these lecture notes the Institute will achieve, in part, its objective of promoting research in the mathematical sciences and their applications.

April 2007

Louis H. Y. Chen
Ka Hin Leung
Series Editors

PREFACE

The primary interest in Gabor and wavelet analyses and their theories stems from and is propelled by the explosion of data arising from rapid advances in communication, sensing and computational power. The usefulness of these data for human knowledge is determined by their accessibility and portability. Pertinent research efforts in mathematics focus on developing new theories, technologies and algorithms for the representation, processing, analysis and interpretation of such scientific data sets. The Gabor and wavelet representations are among the most successful mathematical tools to this end, and were found widespread in applications on signal analysis, image processing and many other information-related areas. Both methodologies deliver representations that are simultaneously local in time and in frequency. The Gabor representation is obtained by windowing the signal through a fixed-size window, which implies that it tiles uniformly the frequency domain. The wavelet representation employs windows that are arbitrarily small, hence is especially suitable for the analysis of transient events.

In response to the recent exciting developments in the mathematical research on image, signal and information processing, the program "Mathematics and Computation in Imaging Science and Information Processing" was held in Singapore at the Institute for Mathematical Sciences, National University of Singapore, from July to December 2003 and in August 2004. A major goal of the program was to promote multidisciplinary research in the area. Among the core topics, one finds time-frequency analysis and applications, wavelet theory and its applications in image and signal processing, and numerical methods in image and information processing. Altogether, three conferences, six workshops, eleven tutorials and three public lectures were set. More than 340 participants took part in these activities including 130 international attendees. We take this opportunity to acknowledge with thanks the essential contribution of the Institute for Mathematical Sciences, which provided us with ample funds and efficient administrative support. We would also like to express our appreciation to the authors for their contributions towards this volume.

The tutorials of the program, each comprising a series of lectures, were conducted by international experts, and they covered a wide spectrum of topics in the field of mathematical image, signal and information processing. The compiled volume includes exposition articles by the tutorial speakers on the foundations of Gabor analysis, subband filters and wavelet algorithms, and operator-theoretic interpolation of wavelets and frames. The volume also presents research papers on Gabor analysis. The accompanying volume focuses more on applications and contains survey articles by the tutorial speakers on subdivision in geometric modeling and computer graphics, high order numerical methods for time dependent Hamilton-Jacobi equations, variational methods in mathematical image processing, data hiding and image steganography, and the apriori algorithm in data mining. The two volumes collectively provide graduate students and researchers new to the field a comprehensive introduction to a number of important topics in mathematical image, signal and information processing. The chapters in each volume were written by specialists in their respective areas. In what follows, we outline the organization of this volume and briefly discuss the topics that are presented.

The first three chapters focus on Gabor frames and its general theory. The first chapter by H. G. Feichtinger, F. Luef and T. Werther is a panoramic introduction to Gabor analysis, from basic principles of linear algebra up to the foundations of advanced time-frequency concepts. It provides the motivation and technical background on some of the key topics in Gabor analysis, namely, the Gabor frame operator, the notion of (dual) Gabor frames, the Janssen representation of the Gabor frame operator, and the spreading function as a tool to describe operators. The first part of the article is devoted to finite dimensional signal spaces, and allows one to get familiar with the general concepts in a concrete linear algebraic setup. The second part deals with Gabor analysis of functions on Euclidean spaces. In this context, the Banach Gelfand triple built upon the Segal algebra appears as a universal tool for a treatment of questions of time-frequency analysis, and the associated modulation spaces.

The second chapter by A. J. E. M. Janssen analyzes several iterative algorithms for computing canonical tight windows and dual windows for Gabor frames by using the calculus of frame operators, the spectral mapping theorem, and Kantorovich's inequality as tools. For the computation of canonical tight windows, both algorithms that require inversion of intermediate frame operators and algorithms that do not need inversions are considered. For the computation of dual windows, algorithms requiring no frame operator inversions are presented. Analysis of the convergence orders of the algorithms is given. The optimality of the convergence orders and the stability of the algorithms are discussed as well.

The third chapter by F. Luef connects Gabor analysis with noncommutative tori and Feichtinger's algebra. In particular, the strong Morita equivalence of noncommutative tori appears as underlying setting for Gabor analysis, since the construction of equivalence bimodules for noncommutative tori has a natural formulation in the notions of Gabor analysis. This leads to the conclusion that Feichtinger's algebra is such an equivalence bimodule. In addition, based on results about Morita equivalence, the biorthogonality relation of Wexler-Raz on the existence of dual atoms of a Gabor frame operator is discussed.

The remaining two chapters are on general wavelet theory. The fourth chapter by P. E. T. Jorgensen expands on some interconnections between mathematical aspects of wavelets and other areas, both within and outside mathematics, such as operator theory, quantum theory, and especially signal processing. Particular attention is given to the concepts of high-pass and low-pass filters: these are central notions in wavelet representation, that played a significant role in signal processing long before wavelets were formally introduced. The chapter gives a detailed account on subband filtering techniques and the efficient algorithms generated by them for practical applications. It also provides connections between these techniques in wavelet analysis and elements of operator theory. It further indicates a number of developments in operator theory which arise from wavelet problems, but are of independent interest in mathematics.

The last chapter by D. R. Larson is an exposition on an operator-theoretic interpolation of wavelets and frames that was developed by the author in collaboration with others. A wavelet is a special case of a vector in a separable Hilbert space that generates a basis under the action of a system of unitary operators. A detailed description of the operator-interpolation approach to wavelet theory using the local commutant of a system is given. This leads to a comprehensive study on applying the theory of operator algebras to wavelet theory. The concrete applications of this method include results obtained using specially constructed families of wavelet sets. The concept of frames is introduced as a sequence of vectors in a Hilbert space which is a compression of a basis for a larger space. Due to this compression relationship between frames and bases, the unitary system approach to wavelets (and more generally, wandering vectors) is perfectly adaptable to frame theory.

Say Song Goh
National University of Singapore, Singapore

Amos Ron
University of Wisconsin-Madison, USA

Zuowei Shen
National University of Singapore, Singapore

A GUIDED TOUR FROM LINEAR ALGEBRA TO THE FOUNDATIONS OF GABOR ANALYSIS

Hans G. Feichtinger, Franz Luef and Tobias Werther

NuHAG, Faculty of Mathematics, University of Vienna
Nordbergstrasse 15, A-1090 Vienna, Austria
E-mail: {hans.feichtinger, franz.luef, tobias.werther}@univie.ac.at

It is the aim of this chapter to provide motivation and technical background on some of the key topics in Gabor analysis: the Gabor frame operator, the notion of (dual) Gabor frames, the Janssen representation of the Gabor frame operator, and the spreading function as a tool to describe operators.

Starting from the context of finite dimensional signal spaces, as it can be constructively realized on a computer and completely described in terms of concepts from linear algebra, we describe first the algebraic side of this problem. The second part of the material emphasizes the necessary definitions and problems (e.g., the form of convergence of various infinite series occurring during the discussion) and provides a description of tools that have become relevant within modern time-frequency analysis in order to overcome the technical problems arising in the context of Gabor analysis involving a continuous domain (such as the Euclidean space). As it is shown, the so-called Banach Gelfand Triple built upon the Segal Algebra $S_0(\mathbb{R}^d)$ appears as the appropriate and universally useful tool for a clear treatment of questions of time-frequency analysis, even if the interest is mostly in operators over L^2. On the other hand a family of function (or distribution) spaces defined by means of time-frequency tools (the so-called modulation spaces) appear to provide a suitable frame-work for a more refined discussion of mathematical problems arising in this field.

1. Introduction

The goal of this report is to provide an introduction to Gabor analysis, from basic principles of linear algebra up to the foundations of advanced time-frequency concepts for Gabor analysis. In order to make the text accessible both to engineers interested in the discrete side of Gabor analysis as well as to mathematicians with a general background in functional analysis there will be essentially two ways to read the article. One path leads directly from linear algebra to "finite Gabor analysis" (more precisely, Gabor analysis for vectors of finite length m, which however are interpreted as periodic sequences, resp. functions over the group of unit roots of order m), and exploits the purely algebraic structure of Gabor systems. This rich structure (e.g., the commutation relation for the frame matrix) implies the possibility to efficiently compute the dual Gabor atom, respectively the bi-orthogonal system, for a linearly independent Gabor family. Note that it is not far from this problem to the discussion of applications of Gabor analysis in the framework of wireless communication (OFDM [45] etc.). On the other hand the numerical realization of the corresponding steps already leads to discussions about the size of certain constants (i.e., the condition number of the frame operator) which may be large (but still finite) in the finite dimensional setting, but could be infinite in similar situations for functions on the real line. Obviously, there the natural domain for Gabor analysis are the square integrable functions, i.e., the Hilbert space $L^2(\mathbb{R})$ respectively in several dimensions $L^2(\mathbb{R}^d)$.

In this context we are facing three additional aspects. The first one is the infinite-dimensionality of the underlying space, and the necessary distinction between bounded and unbounded linear operators. As a consequence there are injective mappings (such as the coefficient mapping with respect to some family of elements in a Hilbert space) which do *not* have a closed range, for example. Indeed, the Gabor system as suggested by D. Gabor in 1946 is such an example (of a total system of vectors, which is *not a frame*), see [21]. These two possible shortcomings of a "generating system" result in the need to require the existence of two constants (essentially equivalent to the boundedness of the frame operator and its inverse). These requirements lead to the concepts of a *frame* and *Riesz bases* respectively, which however are by now quite well understood and can be handled using standard functional analytic concepts.

The second difficulty stems from the fact that - unlike the continuously defined short time Fourier transform (STFT) - a sampled STFT (the starting point of Gabor analysis) for general L^2-windows comes in conflict with the (boundedness) requirements coming in naturally on functional analytic reasons as discussed above. Moreover, one expects in a continuous domain that dual Gabor windows depend continuously on the Gabor atom and the lattice constants in use, and also that the sum describing the frame operator in the Janssen representation converges absolutely. All this can be granted by assuming that atoms are exclusively taken from the Segal Algebra $S_0(\mathbb{R}^d)$, also called Feichtinger's algebra in the literature. It may be not surprising that this space in turn can be defined by means of the STFT, and that it is characterized in the context of Gabor analysis as the collection of those continuous and integrable functions which have - with respect to almost any "nice" Gabor frame - an absolutely convergent Gabor expansion. We will devote some paragraph to the discussions of this space and its properties, but even more so to point out how useful it is in the context of Gabor analysis (and time-frequency analysis in general). It shares many essential properties with the more widely known Schwartz space of "rapidly decreasing functions". Among others it is Fourier invariant, and hence its dual space is a natural domain for an extended Fourier transform. On the other hand it is a simple Banach space with respect to a very natural norm and provides an answer to the problems mentioned above (and many more). The "simple distributions" (one might call the elements of $S_0(\mathbb{R}^d)$ by this name) are however good enough to prove a kernel theorem and to establish a so-called Banach Gelfand Triple (BGT), consisting of the Banach space of test functions $S_0(\mathbb{R}^d)$, the (usual) Hilbert space $L^2(\mathbb{R}^d)$ and the (simple) distribution space $S_0'(\mathbb{R}^d)$. It is quite intuitive to think of these three layers "melted together" in the finite dimensional case, but essentially different in the continuous domain. For example, one can describe the Fourier transform on $\mathbb{R}^d$ as a (unitary) Gelfand triple isomorphism, with the nice fact that the ordinary integral representation of the Fourier transform makes sense (and also the inversion theorem in the pointwise sense) for functions from $S_0(\mathbb{R}^d)$. On the other hand, its behavior at the level of the Hilbert space allows to express the fact that it is unitary, hence preserving angles and L^2-norms. Finally one can claim (at the distributional level) that the Fourier transform maps "pure frequencies" onto the corresponding Dirac-measures (which take of course the role of unit vectors), and that the Fourier transform is uniquely determined (as Gelfand triple isomorphism) by this fact. There are more such Gelfand triple isomorphisms, but we will exploit only the "spreading mapping" a bit more closely, in order to describe the fact that every linear mapping (from S_0 to S_0') is a superposition of TF-shift operators (in a certain sense), while any Gabor frame operator is a (discrete) sum of TF-shifts (the abstract version of Janssen's theorem). In this context we will heavily make use of the so called fundamental relation for Gabor Analysis (FIGA).

The third new point arising in the context of the real line (as opposed to the finite dimensional situation, where any two norms are equivalent) is the need to make a distinction between "good" and "nasty" functions, e.g. using notations related to summability (L^p-spaces in the classical setting of Fourier analysis), smoothness or decay at infinity. One expects that a good frame consisting of "nice atoms" has the property that smooth functions with strong decay at infinity require only a small number of non-zero terms for a good approximation. This is perfectly realized for the classical Besov – Triebel – Lizorkin spaces with respect to "good" wavelet systems and the complete characterization of those spaces via wavelet coefficients is an important aspect of modern wavelet theory. In turn, the "correct" function spaces in the time-frequency context are modulation spaces, introduced in the early eighties. Since they are exactly the spaces of distributions with Gabor coefficients in ℓ^p, the modulation spaces $M^0_{p,p}$ play an important role in time-frequency analysis. As with the case of the classical theory of L^p-spaces the three special cases $p = 1, 2, \infty$ are the most interesting ones, and can be described with less technical effort compared to the more general, by now "classical modulation spaces" $M^s_{p,q}$, which have been modeled in analogy to standard Besov spaces. Since $M^0_{1,1}$ is just the Segal algebra S_0 and its dual $M^0_{\infty,\infty}$ coincides with S_0', we will restrict our attention to these spaces, which make up a *Banach Gelfand Triple* together with $M^0_{2,2} = L^2$. Let us mention however here that this restriction was chosen in order to limit the level of technicality of our presentation, although many of the results involving these spaces carry over to the more general modulation spaces (derived from general solid and translation invariant spaces over phase space) as described by the general coorbit theory developed by Feichtinger and Gröchenig in the late eighties [13].

2. Basics in Linear Algebra

Assume that a signal f is a (column) vector of m complex numbers. Given a family of (column) vectors $g_1, \ldots, g_n$ in $\mathbb{C}^m$ we are looking for a linear combination of these vectors that reproduces f or comes close to f. In other words, we are looking for a coefficient vector c in $\mathbb{C}^n$ such that

$$f \sim c(1)g_1 + \cdots + c(n)g_n \,. \tag{1}$$

We can transfer this problem to matrix notation. Let A be the matrix of size $m \times n$ whose j-th column is g_j. Then equation (1) boils down to the matrix vector product

$$f \sim Ac \,.$$

In the context of the linear mapping $c \mapsto Ac$ induced by A from $\mathbb{C}^n$ (the coefficient space) to $\mathbb{C}^m$ (the signal space), there exist coefficient vectors exactly reproducing f if and only if f is in the range of A, write $R(A)$, which obviously coincides with the linear span of $g_1, \ldots, g_n$ respectively the column-space of A.

So far we have defined the problem and built a simple mathematical model. In order to obtain a better understanding of our problem we need a few basic facts from linear algebra.

A family of vectors $g_1, \ldots, g_n$ is said to be *linearly independent* if

$$a(1)g_1 + \cdots + a(n)g_n = 0$$

implies $a(1) = \ldots = a(n) = 0$. Otherwise the family is *linearly dependent.* Obviously linear independence is another way of expressing the injectivity of the linear mapping induced by A. The *rank* of a matrix A, written as $r(A)$, is the maximal number of linearly independent rows. Important properties are:

(1) $r(A) = r(A')$ where A' denotes the *transpose conjugate* of A.
(2) $r(A)$ coincides with the dimension of $R(A)$.
(3) A of size $m \times n$ is *surjective*, i.e., $R(A) = \mathbb{C}^m$, if and only if $r(A) = m$.

We see that imposing $r(A) = m$ requires $n \geq m$ which is always the case of frames discussed below. In this case, for *every* $f \in \mathbb{C}^m$, there exists at least one coefficient vector c such that $Ac = f$. We are interested in how to find such a coefficient vector.

An easy exercise shows that the nullspaces of A and AA' coincide and therefore A has maximal rank $r(A) = m$ if and only if AA' is invertible. For $r(A) = m$ it follows that $A'(AA')^{-1}$ is well-defined and we can see that

$$c = A'(AA')^{-1}f$$

provides a suitable coefficient vector for reproducing f.

The matrix $A'(AA')^{-1}$ is known as the *pseudo-inverse* of A (whose size is $n \times m$) [24]. Every matrix has a pseudo-inverse which can be computed by the singular value decomposition. In the singular value decomposition the matrix A is decomposed as

$$A = U\Sigma V' ,$$

where the (first $r(A)$) columns of the matrix U are an orthonormal system for the range of A, Σ is a diagonal matrix with the singular values of A (i.e., the eigenvalues of AA') in its diagonal, and V is a well-chosen orthonormal system of $\mathbb{C}^n$. An excellent introduction for the singular value decomposition can be found in [4]. The standard approach for computing the singular value decomposition of a $m \times n$ matrix A ($m \leq n$) is as follows:

(1) Compute the eigenvalues of $AA' = V\Lambda V'$.
(2) Let Σ be the $m \times n$ nonnegative square root of Λ.
(3) Solve the system $U\Sigma = AV$ for unitary U (e.g., via QR-factorization).

The above case with $r(A) = m$ and $m \leq n$ is just a special case for computing a pseudo-inverse.

The question about the uniqueness of the coefficient vector c can be answered as follows. Whenever $n > m$, the family $g_1, \ldots, g_n$ is linearly dependent and any g can be written as a linear combination of the others. Therefore we could, for instance, change one single coefficient and adapt the others such that we recover f. Hence, c is not unique. The coefficients obtained via the pseudo-inverse are of minimal ℓ^2-norm (minimal length). In other words, the pseudo-inverse, applied to the right hand side of a linear problem of the form $Ac = f$ provides the minimal least square solution to this problem.

On the other hand, if $m = n$, then A (assumed to be surjective) is also injective and the difference between c and any other coefficient vector reproducing f must be in the kernel of A which contains only the zero vector. Hence c is unique. Equivalently we could argue that if $m = n$ then the family $g_1, \ldots, g_n$ is linearly independent and, thus, it constitutes a basis for $R(A)$ allowing only unique representations. In the latter case, the pseudo-inverse turns into the standard inverse of A.

Let us come back to the matrix product $A'(AA')^{-1}$. If we consider the complex conjugate of the rows of the pseudo-inverse $A'(AA')^{-1}$ and write them as column vectors, we obtain another family of n vectors in $\mathbb{C}^m$, say $\tilde{g}_1, \ldots, \tilde{g}_n$. As a consequence, we have for all $f \in \mathbb{C}^m$

$$f = AA'(AA')^{-1}f = \sum_{j=1}^{n} \langle f, \tilde{g}_j \rangle g_j \,, \tag{2}$$

where $\langle \cdot, \cdot \rangle$ denotes the standard inner product in $\mathbb{C}^n$. By interchanging the roles of $g_1, \ldots, g_n$ and $\tilde{g}_1, \ldots, \tilde{g}_n$, a simple argument yields

$$f = (AA')^{-1}AA'f = \sum_{j=1}^{n} \langle f, g_j \rangle \tilde{g}_j \,. \tag{3}$$

This shows that the families $\{g_j\}$ and $\{\tilde{g}_j\}$ are dual to each other in the sense that one system provides a left inverse of the other. In the case $m = n$ this is just the dual basis, which in the case of an orthonormal basis coincides with the original system, which amounts to A being unitary.

Besides decomposing a signal, the question of recovering a signal from a coefficient vector is of equal interest in application. That is, given coefficients of the form $\langle f, g_j \rangle$ of a signal f with respect to some basic family $\{g_j\}$, we want to recover the signal f. Fortunately, this can be treated as the dual problem of finding the right coefficients for the corresponding representation. Indeed, in the case where the analyzing family $\{g_j\}$ constitutes a frame, we recover the signal by using a dual frame $\{\tilde{g}_j\}$ as shown in (3).

For practical applications, the numerical properties of the analysis and synthesis mapping are of great importance. These are reflected in the singular values of the matrix A. The singular values coincide with the square root of the eigenvalues of the positive matrix AA'. Specifically, we are interested in the quotient between the maximal $a_{\max}$ and the minimal singular value $a_{\min}$. This quotient is the *condition number* of A and indicates the computational goodness of the algebraic system $Ac = f$.

Let us consider, for example, the family $e_1, e_1 + 10^{-k}e_2$, (e_1, e_2 denote the standard unit vectors in $\mathbb{C}^2$) which is obviously linearly independent in $\mathbb{C}^2$. For constructing e_2, the length of the coefficients is more than 10^k times larger compared to the length of the coefficients for the standard basis. This family is rather badly conditioned.

The minimal and maximal singular value of A are the optimal constants in the following inequality:

$$a_{\min}\|f\|^2 \leq \langle AA'f, f\rangle = \sum_{j=1}^{n} |\langle f, g_j\rangle|^2 \leq a_{\max}\|f\|^2 \tag{4}$$

for all $f \in \mathbb{C}^m$. Inequality (4) is the so-called *frame inequality* (stated in the finite setting). Any family $g_1, \ldots, g_n$ satisfying (4) is called a *frame* for $\mathbb{C}^m$ and the lower and upper bounds are the corresponding frame bounds. This concept will play a crucial role in the continuous "time"-variables.

In the case of $a_{\min} = a_{\max}$, the family $g_1, \ldots, g_n$ is called a *tight frame*, which behaves almost like an orthogonal basis though it might be linearly dependent. Every frame $g_1, \ldots, g_n$ can be converted into the tight frame $S^{-1/2}g_1, \ldots, S^{-1/2}g_n$, where $S^{-1/2}$ is the square root of the inverse of the so-called *frame operator* $S = AA'$. This square root can be obtained via the singular value decomposition of S by simply replacing the singular values by their inverse square root. A simple computation shows that for $A = U\Sigma V'$, the columns of the matrix UV' constitute the above described tight frame $S^{-1/2}g_1, \ldots, S^{-1/2}g_n$.

All these objects have an analogue description for continuous signals in L^2 as described in Section 4. But before, we discuss a particular choice of a basic system that enjoys a lot of structure which in turn induces fast numerical algorithms. In order to better understand this structure we have to describe the theory in a more abstract setting, which requires an extension of the underlying mathematical objects based on classical results in functional analysis.

3. Finite Dimensional Gabor Analysis

In the following we build a special family of basic vectors in $\mathbb{C}^m$ in order to decompose and recover signals of length m as described in the previous section. To this end we essentially take a single vector and derive the remaining basic vectors by regular cyclic shifts and modulations of this vector. In this way, we obtain a highly structured system.

First we introduce the basic notations. For convenience, a vector f in $\mathbb{C}^m$ is periodically extended on $\mathbb{Z}$ via

$$f(k + qm) = f(k), \qquad q \in \mathbb{Z}.$$

The index set of a vector can be identified with the finite group $Z_m =$

$\{1, \ldots, m\}$ and all indices exceeding m are understood modulo m (without explicit notation). In this sense, cyclic shifts are just another form of "ordinary shifts" on periodic functions.

The (index) *translation operator* T_k that rotates the index of a vector by k and the (frequency) *modulation operator* M_l that performs a frequency shift by l have the following matrix entries with respect to the standard basis in $\mathbb{C}^m$:

$$(T_k)_{uv} = \delta_{u+k,v} \quad \text{and} \quad (M_l)_{uv} = e^{2\pi i(u-1)l/m}\delta_{u,v}\,,$$

where δ denotes the Kronecker symbol.

By simply applying the product of these matrices to an arbitrary vector $f = (f(1), \ldots, f(m))^t$ we see that they do not commute in general:

$$M_l T_k f = \left(f(k+1), \ldots, w_m^{(m-k)l} f(m), w_m^{(m-k+1)l} f(1), \ldots, w_m^{(m-1)l} f(k)\right)^t,$$

$$T_k M_l f = \left(w_m^{kl} f(k+1), \ldots, w_m^{(m-1)l} f(m), f(1), \ldots, w_m^{(k-1)l} f(k)\right)^t,$$

where $w_m = e^{2\pi i/m}$. We use the symbol ^{t} to indicate column vectors.

For a more compact description, we write for the time-frequency shift (first a time-shift is applied, and then the modulation corresponding to frequency shift):

$$\pi(\lambda) = M_l T_k \quad \text{with} \quad \lambda = (k, l) \in Z_m \times Z_m\,.$$

Note that $\pi(\lambda)$ are unitary matrices, i.e., $\pi(\lambda)^{-1} = \pi(\lambda)'$. From the relation between $M_l T_k$ and $T_k M_l$, we can easily derive the commutation rule

$$\pi(\lambda_1)\pi(\lambda_2) = e^{2\pi i(l_1 k_2 - k_1 l_2)}\pi(\lambda_2)\pi(\lambda_1)\,. \tag{5}$$

Because of (5), the set of time-frequency shift matrices do not constitute a (multiplicative) group. However, by taking into account this commutation property, time-frequency shift operations can be extended in such a way that they finally constitute a group, namely the so-called *Heisenberg group* [26].

Next we introduce the announced basic system whose structure is heavily based on this commutation rule. Given a lattice (subgroup) Λ in $Z_m \times Z_m$ and a vector g in $\mathbb{C}^m$, we say that the family $\{g_\lambda\}_{\lambda\in\Lambda}$ defined by

$$g_\lambda = \pi(\lambda)g, \qquad \lambda \in \Lambda,$$

is a (discrete) Gabor frame if it spans all $\mathbb{C}^m$. The associated frame matrix is the positive definite matrix S given by

$$Sf = GG'f = \sum_{\lambda\in\Lambda} \langle f, g_\lambda \rangle g_\lambda \,.$$

Every Gabor frame induces a dual frame as described in the previous section. For Gabor frames, however, the computation of the canonical dual can be reduced to the solution of a single linear equation of the form

$$Sh \;=\; g \tag{6}$$

because then the dual frame is simply given by $\{\pi(\lambda)h\}_{\lambda\in\Lambda}$ which makes Gabor frames so attractive for applications. This statement follows easily from the observation that the Gabor frame matrix commutes with all time-frequency shifts $\pi(\lambda)$, $\lambda \in \Lambda$, (and so does S^{-1}) since Λ is supposed to be a group and the factor in (5) simply drops out. What we obtain is the following Gabor representation of a signal $f \in \mathbb{C}^m$:

$$f \;=\; \sum_{\lambda\in\Lambda} \langle f, \pi(\lambda)h \rangle \pi(\lambda) g \;=\; \sum_{\lambda\in\Lambda} \langle f, \pi(\lambda)g \rangle \pi(\lambda) h \,. \tag{7}$$

The Gabor representation shows also how we can reconstruct a signal from the coefficient vector $\langle f, \pi(\lambda)g \rangle$, $\lambda \in \Lambda$, which corresponds to the short-time Fourier transform (STFT) $V_g f$ of f with respect to the window g,

$$V_g f(k,l) = \langle f, \pi(k,l)g \rangle = \sum_{j=1}^{m} f(j)\overline{g(j+k)}e^{-2\pi ijl/m} \,,$$

sampled on the lattice Λ. This leads to the dual perspective in which we ask how we can recover f from samples of its STFT. As seen in the previous section, the frame approach of (7) already gives a satisfactory answer to the dual problem.

Frames, and in particular Gabor frames, are in general linearly dependent, which basically means that not all coefficient information is needed for completely recovering the signal. Indeed, if, for example, one sample of the sampled short-time Fourier transform is missing, say the one indexed by λ_o, we can still reconstruct f since $\{\pi(\lambda)g\}_{\lambda\in\lambda\in\Lambda\setminus\lambda_o}$ is a frame because of the underlying group structure. In this case, however, the dual frame can not be expected to be derived from a single vector.

Equation (6) reveals that one of the most important objects in Gabor analysis is the frame matrix S. Indeed, many studies in this field have been devoted to this operator not only for a better understanding of Gabor systems but also for designing fast numerical inversion schemes. In the last part of this section we give a small insight to fundamental results of the Gabor frame matrix.

We start with the special case of so-called separable lattices of the form $\Lambda = \alpha Z_m \times \beta Z_m$ where α and β are divisors of m. We define $\tilde{\alpha} = m/\alpha$ and $\tilde{\beta} = m/\beta$. (The case that $\alpha\beta$ divides m corresponds to *integer over-sampling.*) Due to the fact that $\sum_{m=0}^{\tilde{\beta}-1} e^{2\pi i j m\beta/m} = 0$ if j does not divide $\tilde{\beta}$, the jl-th element of the frame matrix S is simply given by

$$(S)_{jl} = \begin{cases} \tilde{\beta}\sum_{n=0}^{\tilde{\alpha}-1} g(j-\alpha n)\overline{g(l-\alpha n)} & \text{if } |j-l| \text{ is divided by } \tilde{\beta} \\ 0 & \text{otherwise}, \end{cases} \tag{8}$$

which is called the *Walnut representation* of S for the discrete case [37]. The discrete Walnut representation implies the following properties of S:

(1) Only every $\tilde{\beta}$-th subdiagonal of S is non-zero.
(2) Entries along a subdiagonal are α-periodic.
(3) S is a block circulant matrix of the form

$$S = \begin{bmatrix} A_0 & A_1 & \dots & A_{\tilde{\alpha}-1} \\ A_{\tilde{\alpha}-1} & A_0 & \dots & A_{\tilde{\alpha}-2} \\ \vdots & \vdots & \ddots & \vdots \\ A_1 & A_2 & \dots & A_0 \end{bmatrix}$$

where A_s are non-circulant $\alpha \times \alpha$ matrices, with

$$(A_s)_{j,l} = (S)_{j+s\alpha,l+s\alpha} \tag{9}$$

for $s = 0, 1, \dots, \tilde{\alpha} - 1$ and $j, l = 0, 1, \dots, \alpha - 1$, [42].

This special case applies merely for separable lattices. In general, however, we have another powerful representation of the frame matrix, the so-called *Janssen representation.*

For a better understanding of the Janssen representation in finite dimension we introduce the Frobenius norm for $m \times n$ matrices

$$\|A\|_{\text{Fro}} = \Big(\sum_{i=1}^{m}\sum_{j=1}^{n} |a_{ij}|^2\Big)^{1/2} = \sqrt{\text{tr}(A'A)}$$

where the *trace* $\text{tr}(B)$ of B is the sum of its diagonal entries. (The Frobenius norm corresponds to the Hilbert-Schmidt norm in infinite dimension.) The

Frobenius norm can be derived from the inner product

$$\langle A, B\rangle_{\text{Fro}} = \sum_{i=1}^{m}\sum_{j=1}^{n} a_{ij}\bar{b}_{ij}\,. \tag{10}$$

It can easily be seen that the matrix family $\{\pi(\lambda)\}_{\lambda\in Z_m\times Z_m}$ of all possible time-frequency shifts on $\mathbb{C}^m$ constitutes an orthogonal system with respect to $\langle\cdot,\cdot\rangle_{\text{Fro}}$ in the space of all complex valued $m \times m$ matrices, denoted by $\mathcal{M}(m)$. The system becomes orthonormal for the inner product

$$\langle A, B\rangle_F = \frac{1}{m}\langle A, B\rangle_{\text{Fro}}\,.$$

As a consequence, every matrix in $(\mathcal{M}(m), \langle\cdot,\cdot\rangle_F)$ has a unique expansion (time-frequency representation) with respect to the orthonormal system $\{\pi(\lambda)\}_{\lambda\in Z_m\times Z_m}$ (the coefficients corresponding to the spreading function).

Also the frame matrix S of a Gabor frame can be decomposed with respect to all time-frequency shift matrices. This representation can be simplified due to the special structure of S as we will see in the following.

We define the *adjoint lattice* Λ° of Λ to consist of all tuples of elements in $Z_m \times Z_m$ for which

$$\pi(\lambda)\pi(\lambda^\circ) \;=\; \pi(\lambda^\circ)\pi(\lambda) \tag{11}$$

holds. Remark that Λ° is indeed an (additive) subgroup (lattice) of Z_m. In particulár, according to (5), we must have

$$e^{2\pi i(kl^\circ - k^\circ l)/m} \;=\; 1 \tag{12}$$

for all $\lambda = (k, l)$ and $\lambda^\circ = (k^\circ, l^\circ)$.

We observe that the Gabor frame matrix S commutes with all time-frequency shifts $\pi(\lambda)$ with $\lambda \in \Lambda$. Combining the commutation rule (5) with the uniqueness of the time-frequency representation immediately induces, that the only coefficients that might be different from zero are those related to the dual lattice Λ°. Therefore, we have

$$S \;=\; \sum_{\lambda^\circ\in\Lambda^\circ} c_{\lambda^\circ}\pi(\lambda^\circ) \tag{13}$$

which is the so-called Janssen representation of S. The coefficients are actually given by the restriction of the short time Fourier transform to the dual lattice, i.e.,

$$c_{\lambda^\circ} \;=\; \langle S, \pi(\lambda^\circ)\rangle_F \;=\; \frac{n}{m}\langle g, \pi(\lambda^\circ)g\rangle \;=\; \frac{n}{m}V_g g(\lambda^\circ) \tag{14}$$

where n denotes the order of Λ. In order to derive the relation between the coefficients in (13) and the (discrete) STFT $V_g(g)$ we note that

$$(S)_{jk} = (GG')_{jk} = \sum_{\lambda \in \Lambda} e^{2\pi i (j-k) s_\lambda / m} g(j + r_\lambda) \overline{g}(k + r_\lambda)$$

where $\lambda = (r_\lambda, s_\lambda)$ and

$$(\pi(\lambda^\circ))_{jk} = e^{2\pi i (j-1) s^\circ / m} \delta_{j+r^\circ, k} \,.$$

Now we compute

$$\langle S, \pi(\lambda^\circ) \rangle_F = \frac{1}{m} \sum_{j=1}^{m} \sum_{k=1}^{m} (S)_{jk} \overline{(\pi(\lambda^\circ))}_{jk} =$$

$$\frac{1}{m} \sum_{j=1}^{m} \sum_{k=1}^{m} \Big(\sum_{\lambda \in \Lambda} e^{2\pi i (j-k) s_\lambda / m} g(j + r_\lambda) \overline{g}(k + r_\lambda) \Big) e^{-2\pi i (j-1) s^\circ / m} \delta_{j+r^\circ, k} =$$

$$\frac{1}{m} \sum_{j=1}^{m} \sum_{\lambda \in \Lambda} e^{-2\pi i r^\circ s_\lambda / m} g(j + r_\lambda) \overline{g}(j + r^\circ + r_\lambda) e^{-2\pi i (j-1) s^\circ / m} =$$

$$\frac{1}{m} \sum_{\lambda \in \Lambda} \underbrace{e^{2\pi i (r_\lambda s^\circ - r^\circ s_\lambda)/m}}_{\overset{(12)}{=} 1} \sum_{j=1}^{m} g(j) \overline{g}(j + r^\circ) e^{-2\pi i (j-1) s^\circ / m} =$$

$$\frac{n}{m} \sum_{j=1}^{m} g(j) \overline{(\pi(\lambda^\circ) g)}(j) = \frac{n}{m} \langle g, \pi(\lambda^\circ) g \rangle \,.$$

The coefficients $(c_{\lambda^\circ})_{\lambda^\circ \in \Lambda^\circ}$ are called the *Janssen coefficients* of S.

Relation (14) gives rise to efficient computations of the so-called Janssen coefficients by using the standard fast Fourier transform for the STFT.

The coefficients of the time-frequency expansion correspond to the so-called *spreading function* which will be introduced later in the continuous framework and has revealed many new insights into the study of Gabor analysis.

The ratio n/m denotes the redundancy of the Gabor frame. In case of maximal redundancy m, i.e., $\Lambda = Z_m \times Z_m$, the Gabor frame operator reduces to a multiple of the identity, according to the Janssen representation, and corresponds to a tight frame in which the synthesis Gabor atom coincides with the canonical analysis atom. This situation simply reflects the full STFT and its inversion formula in analogy to the continuous case.

Having two Gabor systems $\{g_\lambda\}$ and $\{\gamma_\lambda\}$, the corresponding frame type operator $S_{\gamma,g} = \Gamma G'$ does also commute with all $\pi(\lambda)$ and we again have

$$S_{\gamma,g} = \sum_{\lambda^\circ \in \Lambda^\circ} c_{\lambda^\circ} \pi(\lambda^\circ)$$

with $c_{\lambda^\circ} = \frac{n}{m} V_\gamma(g)(\lambda^\circ) = \frac{n}{m}\langle \gamma, \pi(\lambda^\circ) g\rangle$ which implies that g and γ are dual frames if and only if $\langle \gamma, \pi(\lambda^\circ) g\rangle = \delta_{0,\lambda^\circ}$, $\lambda^\circ \in \Lambda^\circ$ (*Wexler-Raz identity*).

All the stated results have an analogue description for continuous signals (where the signal space is infinite dimensional). However, convergence problems (not occurring in finite dimension) require a careful treatise of this subject. In our next section we discuss basic functional analytic principles that allow to generalize the Gabor concept to continuous signals mainly by controlling convergence issues.

4. Frames and Riesz Bases

Let $\{g_k\}_{k\in\mathbb{Z}}$ be a family in an infinite dimensional (separable) Hilbert space $\mathcal{H}$ with inner product $\langle \cdot, \cdot \rangle$. The classical examples are the Hilbert space L^2 of (equivalence classes of measurable) functions of finite energy (L^2-norm) and the sequence space ℓ^2 consisting of square summable complex-valued sequences. Similar to the finite dimensional case, we want to represent a signal f in $\mathcal{H}$ as a (now possibly infinite) linear combination of the form

$$f \sim \sum_k c_k g_k \,.$$

At this point there are several questions that arise naturally when considering an infinite sum. First, by convention, we want the sum to converge in the (prescribed) Hilbert-norm, i.e., $\lim_{K\to\infty} \|f - \sum_{k=1}^K c_k g_k\| \longrightarrow 0$. Secondly, the sum should converge to the same limit (preferably f) regardless of the summation order we choose (known as *unconditional convergence*). A more subtle point is that we would like to have a continuous linear dependency between the signal f and the coefficients c_k in order to avoid pathological cases in which small alterations in the signal result in uncontrollable changes in the corresponding coefficient sequence and vice-versa. This technical detail accounts for numerical stability.

Obviously, all these requirements are trivially fulfilled in the finite dimensional case. For an infinite family $\{g_k\}$, however, these assumptions have to be ensured before dealing with decomposition and reconstruction issues. Fortunately, there exist concepts in functional analysis that do exactly fit this kind of requirements. For a precise description we need the following definitions.

Definition 1: A family $\{g_k\}$ of a Hilbert space $\mathcal{H}$ is complete in $\mathcal{H}$ if the set of finite linear combinations of $\{g_k\}$, write $\mathrm{span}(g_k)$, is dense in $\mathcal{H}$, i.e., every f in $\mathcal{H}$ can be arbitrarily well approximated by elements in $\mathrm{span}(g_k)$ with respect to the $\mathcal{H}$-norm.

In the mathematical literature complete systems are often called "total". The definition makes no claim about the "cost" of approximation. In other words, it is allowed to use more and more complicated coefficient sequences as the approximation quality is increased. In particular, total families do not necessarily allow a series expansion of arbitrary elements from the given Hilbert space.

Definition 2: The family $\{g_k\}$ of a Hilbert space $\mathcal{H}$ is a basis for $\mathcal{H}$ if for all $f \in \mathcal{H}$ there exists unique scalars $c_k(f)$ such that

$$f = \sum_k c_k(f) g_k \, .$$

In contrast to complete sequences, a basis always induces a series expansion.

Definition 3: The sequence $\{g_k\}$ in a Hilbert space $\mathcal{H}$ is called a Bessel sequence if

$$\sum_k |\langle f, g_k \rangle|^2 < \infty \, , \qquad f \in \mathcal{H} \, .$$

Definition 4: A family $\{g_k\}$ of a Hilbert space $\mathcal{H}$ is a Riesz sequence if there exist bounds $A, B > 0$ such that

$$A\|c\|_{\ell^2}^2 \leq \left\| \sum_k c_k g_k \right\|^2 \leq B\|c\|_{\ell^2}^2 \, , \qquad c \in \ell^2 \, .$$

Riesz sequences preserve many properties of orthonormal sets [5]. A Riesz sequence which generates all $\mathcal{H}$ is called a *Riesz basis* for $\mathcal{H}$. Riesz bases are somehow "distorted" orthonormal bases as described in the following lemma which reveals all useful properties of a Riesz basis [28].

Lemma 5: *Let $\{g_k\}$ be a sequence in a Hilbert space $\mathcal{H}$. The following are equivalent.*

(1) *$\{g_k\}$ is a Riesz basis for $\mathcal{H}$.*
(2) *$\{g_k\}$ is an unconditional basis for $\mathcal{H}$ and g_k are uniformly bounded.*
(3) *$\{g_k\}$ is a basis for $\mathcal{H}$, and $\sum_k c_k g_k$ converges if and only if $\sum_k |c_k|^2$ converges.*
(4) *There is an equivalent inner product on $\mathcal{H}$ for which $\{g_k\}$ is an orthonormal basis for $\mathcal{H}$.*
(5) *$\{g_k\}$ is a complete Bessel sequence and possesses a bi-orthogonal system $\{h_k\}$ that is also a complete Bessel sequence.*

The last item of the lemma says that there exists a unique sequence $\{h_k\}$ such that $\langle g_k, h_j\rangle = \delta_{kj}$ which, combined with the second statement, induces the representation

$$f = \sum_k \langle f, h_k\rangle g_k = \sum_k \langle f, g_k\rangle h_k\,, \qquad f \in \mathcal{H}\,.$$

Hence, Riesz bases are potential candidates for our purpose of signal representation. We point out that the coefficient sequence is always square summable which is an important stability criterion. In the next section we will give an example of a basis which is not stable, i.e., the coefficient sequence is not summable at all for special examples.

So far, the systems we are considering allow only unique expansions with respect to the coefficients. In applications it is sometimes more useful to weaken this property. This can be obtained by looking for overcomplete (linearly dependent) sets which is implemented in the concept of frames introduced by Duffin and Schaeffer in 1952 [11].

Definition 6: A family $\{g_k\}$ of a Hilbert space $\mathcal{H}$ is a frame of $\mathcal{H}$ if there exist bounds $A, B > 0$ such that

$$A\|f\|^2 \leq \sum_{k\in\mathbb{Z}} |\langle f, g_k\rangle|^2 \leq B\|f\|^2\,, \qquad f \in \mathcal{H}\,. \tag{15}$$

If $A = B$, then $\{g_k\}_{k\in\mathbb{Z}}$ is called a tight frame.

The synthesis map $G : \ell^2 \to \mathcal{H}$ of a frame $\{g_k\}$ is defined by

$$G : (c_k) \to \sum_k c_k g_k \,.$$

Its adjoint G^* is the analysis operator $G^* f = (\langle f, g_k \rangle)$. The *frame operator* S is defined by

$$Sf = GG^* f = \sum_{k \in \mathbb{Z}} \langle f, g_k \rangle g_k \,, \qquad f \in \mathcal{H} \,.$$

By (15), the frame operator satisfies

$$A\langle f, f \rangle \leq \langle Sf, f \rangle \leq B \langle f, f \rangle \,, \qquad f \in \mathcal{H} \,,$$

and is, therefore, bounded, positive, and invertible. The inverse operator S^{-1} is obviously also positive and has therefore a square root $S^{-1/2}$ (self-adjoint), [40]. It follows that the sequence $\{S^{-1/2} g_k\}$ is a tight frame with $A = B = 1$.

Every orthonormal basis of $\mathcal{H}$ is a Riesz basis of $\mathcal{H}$ and every Riesz basis of $\mathcal{H}$ is also a frame. The important difference between a Riesz basis and a frame is that the null space $\mathcal{N}(G)$ of the synthesis map G of a frame $\{g_k\}$ is in general non-trivial, which is equivalent to the statement that the range of the analysis map G^* is a (closed) proper subspace of ℓ^2.

The sequence $\{\tilde{g}_k\}$ with $\tilde{g}_k = S^{-1} g_k$, is also a frame with frame bounds $1/B$ and $1/A$. It is a dual frame for $\{g_k\}$ in the sense that

$$f \;=\; \sum_k \langle f, \tilde{g}_k \rangle g_k \;=\; \sum_k \langle f, g_k \rangle \tilde{g}_k \,, \qquad f \in \mathcal{H} \,.$$

Again, we see that frames do indeed fit our purpose for signal analysis and signal recovery. In contrast to Riesz bases, frames have, in general, no bi-orthogonal relation. Moreover, the dual frame is not unique. The canonical dual $\{S^{-1} g_k\}$ is the one that produces minimal ℓ^2 coefficients as already shown in [11]. It corresponds to the pseudo-inverse of the analysis operator in finite dimension. For alternative dual frames there exist constructive approaches that rely on the canonical dual. In [6,33], it is shown that any dual frame of $\{g_k\}$ can be written as

$$S^{-1} g_k + h_k - \sum_j \langle S^{-1} g_k, g_j \rangle h_j \,, \tag{16}$$

where $\{h_k\}$ is a Bessel sequence.

The lack of uniqueness has the advantage that if one coefficient is missing out of the sequence $\langle f, g_k \rangle$, the whole signal can still be completely recovered as long as $\{g_k\}$ is a frame but no Riesz basis. Similarly, any frame that is not a Riesz basis is still a frame when discarding single frame elements. Studies about the conservation of the frame property when discarding frame elements are known as *excesses of frames* [1,2].

5. Gabor Analysis on L^2

We define the Fourier transform of an integrable function by $\hat{f}(\omega) = \int_{\mathbb{R}^d} f(t) e^{-2\pi i t \omega} dt$. For an element $\lambda = (x, \omega) \in \mathbb{R}^{2d}$ we define the time-frequency shift $\pi(\lambda)$ by

$$\pi(\lambda) = M_\omega T_x$$

where T_x is the translation operator $T_x f = f(\cdot - x)$ and M_ω is the modulation operator $M_\omega f = e^{2\pi i \omega \cdot} f(\cdot)$. In analogy to the finite dimensional model, note that

$$\pi(\lambda_2)\pi(\lambda_1) = e^{2\pi i (x_1 \omega_2 - x_2 \omega_1)} \pi(\lambda_1)\pi(\lambda_2)$$

for $\lambda_1 = (x_1, \omega_1), \lambda_2 = (x_2, \omega_2) \in \mathbb{R}^{2d}$.

A time-frequency lattice Λ is a discrete subgroup of $\mathbb{R}^{2d}$ $(= \mathbb{R}^d \times \hat{\mathbb{R}}^d)$ with compact quotient. Its redundancy $|\Lambda|$ is the reciprocal value of the measure of a fundamental domain for the quotient $\mathbb{R}^{2d}/\Lambda$.

For a lattice Λ in $\mathbb{R}^{2d}$ and a so-called *Gabor atom* $g \in L^2$ we define the associated Gabor family by

$$\mathcal{G}(g, \Lambda) = \{\pi(\lambda) g\}_{\lambda \in \Lambda} .$$

If $\mathcal{G}(g, \Lambda)$ is a frame for L^2, we call it a *Gabor frame*. Since Λ has a group structure, the frame operator

$$Sf = \sum_{\lambda \in \Lambda} \langle f, \pi(\lambda) g \rangle \pi(\lambda) g$$

has the property that it commutes with all time-frequency shifts of the form $\pi(\lambda)$ for $\lambda \in \Lambda$. Therefore, the canonical dual frame of $\mathcal{G}(g, \Lambda)$ is simply given by $\mathcal{G}(h, \Lambda)$ with $h = S^{-1} g$. The fact that a canonical dual frame of a Gabor frame is again a Gabor frame, i.e., generated by a single function, is the key property in many applications. It reduces computational issues to solving the linear system $Sh = g$.

A special and widely studied case is separable lattices of the form $\alpha\mathbb{Z} \times \beta\mathbb{Z}$ for some positive lattice parameters α and β, whose redundancy is simply $(\alpha\beta)^{-1}$. The prototype of a function generating Gabor frames for such separable lattices is the Gaussian

$$\psi(x) = e^{-\pi x^2 \sigma^2} \tag{17}$$

for some real $\sigma > 0$. The Gaussian generates a Gabor frame if and only if $\alpha\beta < 1$ [36,41]. We emphasize that for $\alpha\beta = 1$ the Gaussian generates an *unstable* generating system for L^2, i.e., the resulting Gabor family is complete but coefficient sequences must not be bounded. In this context we mention a central result, the so-called *density theorem* and refer to [26] for detailed discussions. An elegant elementary proof of the density theorem has been provided by Janssen [30].

Theorem 7: *Assume that $\mathcal{G}(g, \alpha, \beta)$ is a frame. Then, $\alpha\beta \leq 1$. Moreover, $\mathcal{G}(g, \alpha, \beta)$ is a Riesz basis for L^2 if and only if $\alpha\beta = 1$.*

In his seminal paper [21], Gabor chose the integer lattice $\alpha = \beta = 1$ in $\mathbb{R}^2$ and used the Gaussian in order to define a Gabor system with maximal time-frequency localization. However, as mentioned above, this system is no longer stable though complete, and, indeed, the celebrated Balian-Low Theorem [3,35] states that good time-frequency localization and Gabor Riesz bases are not compatible:

Theorem 8: *(Balian-Low) If $\mathcal{G}(g, 1, 1)$ constitutes a Riesz basis for $L^2(\mathbb{R})$, then*

$$\int_{\mathbb{R}} |g(t)|^2 t^2 dt \int_{\mathbb{R}} |\hat{g}(\omega)|^2 \omega^2 d\omega = \infty \,.$$

The Balian-Low Theorem reveals a form of uncertainty principle and has inspired fundamental research, see [26] and references therein.

In the sequel we state some fundamental results on Gabor frames and the Gabor frame operator (Gabor frame-type operator). To this end we need the notion of the *adjoint lattice* Λ° of Λ which is, similarly to the discrete case, the set of all elements in $\mathbb{R}^{2d}$ that satisfy the commutation property

$$\pi(\lambda^\circ)\pi(\lambda) = \pi(\lambda)\pi(\lambda^\circ) \quad \text{for all} \quad \lambda \in \Lambda \,.$$

Note that Λ° is again a lattice of $\mathbb{R}^{2d}$ (and that $\Lambda^{\circ\circ} = \Lambda$). Instead of the frame operator we will use the more general notion of a frame-type operator $S_{g,\gamma,\Lambda}$ associated to the pair (g, γ), where γ takes the role of an "analyzing" and g the role of a "synthesizing" window:

$$S_{g,\gamma,\Lambda} f = \sum_{\lambda \in \Lambda} \langle f, \pi(\lambda)\gamma \rangle \, \pi(\lambda) g \,, \qquad f \in L^2 \,.$$

This sum converges in L^2 for all $f \in L^2$ as long as both functions g, γ are Bessel atoms for Λ, that is, $\mathcal{G}(g, \Lambda)$ and $\mathcal{G}(\gamma, \Lambda)$ are Gabor Bessel families. For the fundamental results to hold with respect to norm convergence we need a little bit more than Bessel sequences, namely, that both atoms g and γ satisfy

$$\sum_{\lambda^\circ \in \Lambda^\circ} |\langle g, g_{\lambda^\circ} \rangle| < \infty \,. \tag{A'}$$

A' is also known as the *Tolimieri-Orr's condition.* This somewhat technical property is used for controlling convergence problems (by altering the convergence definition Condition A' can be weakened). Condition A' is in general not easy to verify. In particular, if Condition A' holds for one lattice, there is, in general, no guarantee that it holds also for a different lattice. This problem, however, is overcome by the Feichtinger algebra S_0 which defines a class of functions for which Condition A' is satisfied for any lattice in $\mathbb{R}^{2d}$. We introduce this algebra in Section 7.

We summarize the fundamental results of Gabor analysis in the following theorem that is given in [19] in a slightly more general context. The statements go back to the seminal papers [9,29,43]. They, however, are all consequences of the fundamental identity of Gabor analysis extensively studied in [17].

Theorem 9: *Let Λ be a lattice in $\mathbb{R}^{2d}$ with adjoint lattice Λ°. Then, for g, h satisfying A', the following hold.*

(1) (*Fundamental Identity of Gabor Analysis*)

$$\sum_{\lambda \in \Lambda} \langle f, \pi(\lambda)\gamma \rangle \langle \pi(\lambda) g, h \rangle = |\Lambda| \sum_{\lambda^\circ \in \Lambda^\circ} \langle g, \pi(\lambda^\circ)\gamma \rangle \, \langle \pi(\lambda^\circ) f, h \rangle \tag{18}$$

for all $f, h \in L^2$, where both sides converge absolutely.

(2) (*Wexler-Raz Identity*)

$$S_{g,\gamma,\Lambda} f = |\Lambda| \cdot S_{f,\gamma,\Lambda^\circ} g \tag{19}$$

for all $f \in L^2$.

(3) (*Janssen Representation*)

$$S_{g,\gamma,\Lambda} = |\Lambda| \sum_{\lambda^\circ \in \Lambda^\circ} \langle \gamma, \pi(\lambda^\circ) g \rangle \, \pi(\lambda^\circ) \tag{20}$$

where the series converges unconditionally in the strong operator sense.

In Section 8 we explicitly derive the Janssen representation of the Gabor frame operator from advanced concepts in harmonic analysis and provide a much deeper insight into this topic.

Another important result is the Ron-Shen Duality Principle which is often referred to [39] although it appeared already in [29] and [9] and was announced in [38].

Theorem 10: *Let* $g \in L^2$ *and* Λ *be a lattice in* $\mathbb{R}^{2d}$ *with adjoint* Λ°. *Then the Gabor system* $\mathcal{G}(g, \Lambda)$ *is a frame for* L^2 *if and only if* $\mathcal{G}(g, \Lambda^\circ)$ *is a Riesz basis for its closed linear span. In this case, the quotient of the two frame bounds and quotient of the Riesz bounds (alternatively the condition number of the corresponding frame operator and the Gramian matrix, respectively), coincide.*

The last important identity in Gabor analysis that we want to present in this section is the Wexler-Raz Biorthogonality Relation which basically says that g and γ are dual Gabor windows if and only if $S_{g,\gamma,\Lambda} = Id$. That is, according to Janssen Representation, exactly the case when

$$\langle \gamma, \pi(\lambda^\circ) g \rangle \; = \; |\Lambda|^{-1} \, \delta_{0,\lambda^\circ} \, .$$

Alternatively, this relation can be described by what is a true biorthogonality (using again Kronecker's Delta):

$$\langle \pi(\lambda^{\circ\prime})\gamma, \pi(\lambda^\circ) g \rangle \; = \; |\Lambda|^{-1} \, \delta_{\lambda^{\circ\prime},\lambda^\circ} \, .$$

So far we have seen that, similar to the finite dimensional model, the Gabor frame operator in the continuous case plays a central role in Gabor theory. Indeed, it is the key object that allows for opening different perspectives, and bridges Gabor analysis to other research areas. In the next section we describe basic and more advanced studies in harmonic analysis that contribute to a better understanding of the Gabor frame operator.

6. Time-Frequency Representations

Traditionally we extract the frequency information of a signal f by means of the Fourier transform $\hat{f}(\omega) = \int_{\mathbb{R}^d} f(t)e^{-2\pi it\omega}dt$. If we know $\hat{f}(\omega)$ for all frequencies ω, then our signal f can be reconstructed by the inversion formula $f(t) = \int_{\widehat{\mathbb{R}^d}} \hat{f}(\omega)e^{2\pi it\omega}d\omega$ (valid pointwise or in the quadratic mean).

However, in many situations it is of relevance to know, *how long* each frequency appears in the signal f, e.g., for a pianist playing a piece of music. Mathematically this leads to the study of functions $S(f)(t,\omega)$ of the signal f, which describe the time-frequency content of f over "time" t. In the following we mention the most prominent time-frequency representations.

In the last century researchers such as E. Wigner, Kirkwood, and Rihaczek had invented different time-frequency representations, [44,32]. The work of Wigner and Kirkwood was motivated by the description of a particle in quantum mechanics by a joint probability distribution of position and momentum of the particle. More concretely, in 1932 Wigner introduced the first time-frequency representation of a function $f \in L^2(\mathbb{R}^d)$ by

$$W(f)(x,\omega) = \int_{\mathbb{R}^d} f(x+\frac{t}{2})\overline{f(x-\frac{t}{2})}e^{-2\pi i\omega t}dt,$$

the so called *Wigner distribution* of f. Later Kirkwood proposed another time-frequency representation, which was in a different context rediscovered by Rihaczek. Both researchers associated to a function $f \in L^2(\mathbb{R}^d)$ the following expression

$$R(f)(x,\omega) = f(x)\overline{\hat{f}(\omega)}e^{-2\pi ix\omega},$$

the *Kirkwood-Rihaczek distribution* of f.

Nowadays, the *short-time Fourier transform* (STFT) has become the standard tool for (linear) time-frequency analysis. It is used as a measure of the time-frequency content of a signal f (energy distribution), but it also establishes a connection to the Heisenberg group.

The STFT provides information about local (smoothness) properties of the signal f. This is achieved by localization of f near t through multiplication with some *window function* g and subsequently applying the Fourier transform providing information about the frequency content of f in this segment. Typically g is concentrated around the origin. If g is compactly supported only a segment of f in some interval or ball around t is relevant, but g can be any non-zero Schwartz function such as the Gaussian. Overall we have:

$$V_g f(x,\omega) = \int_{\mathbb{R}^d} f(t)\overline{g(t-x)}e^{-2\pi i t\omega}dt, \quad \text{for} \quad (x,\omega) \in \mathbb{R}^{2d}. \tag{21}$$

Before we discuss the properties of the STFT we recall the key players of our game: *translation* T_x, *modulation* M_ω, and *time-frequency shifts* $\pi(x,\omega) = M_\omega T_x$ of a signal f already introduced in Section 5.

In 1927 Weyl pointed out that the translation and modulation operator satisfy the following commutation relation

$$T_x M_\omega = e^{-2\pi i x\omega} M_\omega T_x, \quad (x,\omega) \in \mathbb{R}^{2d}. \tag{22}$$

$\{T_x : x \in \mathbb{R}^d\}$ and $\{M_\omega : \omega \in \mathbb{R}^d\}$ are Abelian groups of unitary operators, with the infinitesimal generators given by differentiation and multiplication operator, respectively. Therefore the commutation relation (22) is the analogue of Heisenberg's commutation relation for the differentiation and multiplication operator.

The time-frequency shifts $M_\omega T_x$ for $(x,\omega) \in \mathbb{R}^{2d}$ satisfy the following composition law:

$$\pi(x,\omega)\pi(y,\eta) = e^{-2\pi i x\cdot\eta}\pi(x+y,\omega+\eta), \tag{23}$$

for $(x,\omega),(y,\eta)$ in the time-frequency plane $\mathbb{R}^d \times \widehat{\mathbb{R}^d}$. I.e. the mapping $(x,\omega) \mapsto \pi(x,\omega)$ defines (only) a *projective representation* of the time-frequency plane (viewed as an Abelian group) $\mathbb{R}^d \times \widehat{\mathbb{R}^d}$. By adding a toral component, i.e. $\tau \in \mathbb{C}$ with $|\tau| = 1$ one can augment the phase space $\mathbb{R}^d \times \widehat{\mathbb{R}^d}$ to the so-called *Heisenberg group* $\mathbb{R}^d \times \widehat{\mathbb{R}^d} \times T$ and the mapping $(x,\omega,\tau) \mapsto \tau M_\omega T_x$ defines a (true) unitary representation of the Heisenberg group [20], the so-called *Schrödinger representation.* From this point of view the definition of $V_g f$ can be interpreted as representation coefficients:

$$V_g f(x,\omega) = \langle f, M_\omega T_x g\rangle, \qquad f,g \in L^2(\mathbb{R}^d).$$

The STFT is linear in f and conjugate linear in g. The choice of the window function g influences the properties of the STFT remarkably. One example of a good window class is the Schwartz space of rapidly decreasing functions. Later we will discuss another function space, which is perfectly suited as a good class of windows, *Feichtinger's algebra.*

Furthermore, for $f, g \in L^2(\mathbb{R}^d)$ the STFT $V_g f$ is uniformly continuous on $\mathbb{R}^{2d}$, i.e., we can sample the $V_g f$ without a problem. This fact is of great relevance in the discussion of Gabor frames.

By Parseval's theorem and an application of the commutation relations (22) we derive the following relation

$$V_g f(x, \omega) = e^{-2\pi i x \omega} V_{\hat{g}} \hat{f}(\omega, -x), \tag{24}$$

which is sometimes called the *fundamental identity of time-frequency analysis* [26]. The equation (24) expresses the fact that the STFT is a joint time-frequency representation and that the Fourier transform amounts to a rotation of the time-frequency plane $\mathbb{R}^d \times \widehat{\mathbb{R}^d}$ by an angle of $\frac{\pi}{2}$ whenever the window g is Fourier invariant. Another important consequence of the definition of STFT (21) and the commutation relations (22) is the *covariance property* of the STFT:

$$V_g(T_u M_\eta f)(x, \omega) = e^{-2\pi i u \omega} V_g f(x - u, \omega - \eta). \tag{25}$$

Later we will draw an important conclusion of the basic identity of time-frequency analysis (24) and the covariance property of the STFT (25): isometric Fourier invariance and the invariance under TF-shifts of Feichtinger's algebra.

As for the Fourier transform there is also a Parseval's equation for the STFT which is referred to as *Moyal's formula.*

Lemma 11: *(Moyal's Formula) Let $f_1, f_2, g_1, g_2 \in L^2(\mathbb{R}^d)$ then $V_{g_1} f_1$ and $V_{g_2} f_2$ are in $L^2(\mathbb{R}^{2d})$ and the following identity holds:*

$$\langle V_{g_1} f_1, V_{g_2} f_2 \rangle_{L^2(\mathbb{R}^{2d})} = \langle f_1, f_2 \rangle \overline{\langle g_1, g_2 \rangle}. \tag{26}$$

Moyal's formula implies that orthogonality of windows g_1, g_2 resp. of signals f_1, f_2 implies orthogonality of their STFT's. Most importantly, we observe that for normalized $g \in L^2(\mathbb{R}^d)$ (i.e., with $\|g\|_2 = 1$) one has:

$$\|V_g f\|_{L^2(\mathbb{R}^{2d})} = \|f\|_{L^2(\mathbb{R}^d)},$$

for all $f \in L^2(\mathbb{R}^d)$, i.e., the STFT is an isometry from $L^2(\mathbb{R}^d)$ to $L^2(\mathbb{R}^{2d})$.

Another consequence of Moyal's formula is an inversion formula for the STFT. Assume that the analysis window $g \in L^2(\mathbb{R}^d)$ and the synthesis window $\gamma \in L^2(\mathbb{R}^d)$ satisfy $\langle g, \gamma \rangle \neq 0$. Then for $f \in L^2(\mathbb{R}^d)$

$$f = \frac{1}{\langle g, \gamma \rangle} \iint_{\mathbb{R}^{2d}} \langle f, \pi(x, \omega) g \rangle \, \pi(x, \omega) \gamma \, dx d\gamma. \tag{27}$$

We observe that in contrast to the Fourier inversion the building blocks of the STFT inversion formula are just time-frequency shifts of a square-integrable function. Therefore the Riemannian sums corresponding to this inversion integral are functions in $L^2(\mathbb{R}^d)$ and are even norm convergent in $L^2(\mathbb{R}^d)$ for nice windows (from Feichtinger's algebra, see later).

Recently the Heisenberg uncertainty principle, which expresses in many different ways that *a function f and its Fourier transform $\hat{f}$ cannot be both localized simultaneously*, has received much attention. We point the reader to the work of Gröchenig, Janssen, Hogan, Lakey, and others for results on uncertainty principles for time-frequency representations.

A first example is the following result of Lieb [34], which is derived from an application of Beckner's sharp Hausdorff-Young and Young's inequalities to $V_g f(x,\omega) = (\widehat{f \cdot T_x \overline{g}})(\omega)$:

Lemma 12: *If $f, g \in L^2(\mathbb{R}^d)$ and $1 \leq p < 2$, then*

$$\left(\iint_{\mathbb{R}^{2d}} |V_g f(x,\omega)|^p dx d\omega \right)^{1/p} \geq \left(\frac{2}{p} \right)^{d/p} \|f\|_2 \|g\|_2 \,. \tag{28}$$

The inequality is reversed for $2 \leq p < \infty$.

Results of this kind can be taken as a motivation (although it was not the original one) for the introduction of function spaces characterized by summability and decay properties of the STFT of their elements. In 1983 Feichtinger introduced such a family of Banach spaces, the so-called *modulation spaces.* They have shown to be the right setup for a deeper understanding of operators in time-frequency analysis. In the sequel we will meet two members of the scale of modulation spaces: Feichtinger's algebra $S_0(\mathbb{R}^d)$ and its dual space $S_0'(\mathbb{R}^d)$. In this setup Lieb's inequality expresses just embeddings of certain modulation spaces into $L^2(\mathbb{R}^d)$. Gröchenig and some of his collaborators have extensively studied uncertainty principles as embeddings of certain weighted L^p-spaces into modulation spaces, [25].

7. The Gelfand Triple $(S_0, L^2, S_0')(\mathbb{R}^d)$

Since Feichtinger's discovery of the Segal algebra $S_0(\mathbb{R}^d)$ in 1979, many results have shown that $S_0(\mathbb{R}^d)$ is a good substitute for Schwartz's space $\mathcal{S}(\mathbb{R}^d)$ of test functions (except if one is interested in a discussion of partial differential equations). Furthermore, $S_0(\mathbb{R}^d)$ has turned out to be the appropriate setting for the treatment of questions in harmonic analysis on $\mathbb{R}^d$ (actually on a general locally compact Abelian group G, even without using structure theory). In this section we recall well-known properties of $S_0(\mathbb{R}^d)$ which we will need later in our discussion of Gabor frame operators. Nowadays the space $S_0(\mathbb{R}^d)$ is called *Feichtinger's algebra* since it is a Banach algebra with respect to pointwise multiplication and convolution.

A function in $f \in L^2$ is (by definition) in the subspace $S_0(\mathbb{R}^d)$ if for some non-zero g (called the "window") in the Schwartz space $\mathcal{S}(\mathbb{R}^d)$

$$\|f\|_{S_0} := \|V_g f\|_{L^1} = \iint_{\mathbb{R}^d \times \widehat{\mathbb{R}}^d} |V_g f(x,\omega)| dx d\omega < \infty.$$

The space $(S_0(\mathbb{R}^d), \|\cdot\|_{S_0})$ is a Banach space, for any fixed, non-zero $g \in S_0(\mathbb{R}^d)$, and different windows g define the same space and equivalent norms. Since $S_0(\mathbb{R}^d)$ contains the Schwartz space $\mathcal{S}(\mathbb{R}^d)$, any Schwartz function is suitable, but also compactly supported functions having an integrable Fourier transform (such as a trapezoidal or triangular function) are suitable windows. Often it is convenient to use the Gaussian as a window.

The above definition of $S_0(\mathbb{R}^d)$ (different from the original one) allows for an easy derivation of the basic properties of Feichtinger's algebra in the following lemma. Although they have appeared in various publications, cf. [26], we include the proofs as examples for the derivation of norm-estimates, as they are used in many different areas of time-frequency analysis.

Lemma 13: *Let $f \in S_0(\mathbb{R}^d)$, then the following holds:*

(1) $\pi(u,\eta)f \in S_0(\mathbb{R}^d)$ *for* $(u,\eta) \in \mathbb{R}^d \times \widehat{\mathbb{R}}^d$, *and* $\|\pi(u,\eta)f\|_{S_0} = \|f\|_{S_0}$.
(2) $\hat{f} \in S_0(\mathbb{R}^d)$, *and* $\|\hat{f}\|_{S_0} = \|f\|_{S_0}$.

Proof:

(1) For $z = (u,\eta)$ in the time-frequency plane $\mathbb{R}^d \times \widehat{\mathbb{R}}^d$ one has:

$$\|\pi(u,\eta)f\|_{S_0} = \iint_{\mathbb{R}^{2d}} |V_g f(x-u,\omega-\eta)| \, dx d\omega$$

$$= \iint_{\mathbb{R}^{2d}} |V_g f(x,\omega)| \, dx d\omega = C\|f\|_{S_0}.$$

(2) The key of the argument is an application of the fundamental identity of time-frequency analysis (24) to a Fourier invariant window g and the independence of the definition of $S_0(\mathbb{R}^d)$ from $g \in \mathcal{S}(\mathbb{R}^d)$. For simplicity we choose g the (Fourier invariant) Gaussian $g_0(x) = 2^{d/4}e^{-\pi x^2}$:

$$\|\hat{f}\|_{S_0} = \iint_{\mathbb{R}^{2d}} |V_{g_0}\hat{f}(x,\omega)|\,dxd\omega = \iint_{\mathbb{R}^{2d}} |V_{\widehat{g_0}}\hat{f}(x,\omega)|\,dxd\omega =$$

$$\iint_{\mathbb{R}^{2d}} |V_{g_0}f(-\omega,x)|\,dxd\omega = \iint_{\mathbb{R}^{2d}} |V_{g_0}f(x,\omega)|\,dxd\omega = \|f\|_{S_0}. \quad \square$$

Later we will need that $S_0(\mathbb{R}^d)$ is *dense* and *continuously embedded* into $L^p(\mathbb{R}^d)$ for any $p \in [1,\infty)$. The original motivation for Feichtinger's introduction of $S_0(\mathbb{R}^d)$ was the search for a *smallest* member in the family of all time-frequency homogenous Banach spaces. For a proof of all these assertions we refer the reader to the original paper of Feichtinger [12] or Gröchenig's book on time-frequency analysis [26].

Another reason for the usefulness of $S_0(\mathbb{R}^d)$ is the fact that $S_0(\mathbb{R}^d)$ is a natural domain for the application of Poisson summation formula [25].

Lemma 14: *Let Λ be a lattice in $\mathbb{R}^d$ and $f \in S_0(\mathbb{R}^d)$ then*

$$\sum_{\lambda\in\Lambda} f(\lambda) = |\Lambda|^{-1} \sum_{\lambda^\perp\in\Lambda^\perp} \hat{f}(\lambda^\perp) \tag{29}$$

holds pointwise and with absolute convergence.

Here $\Lambda^\perp$ is the orthogonal lattice for Λ, e.g. $\Lambda^\perp = (A^{-1})^t\mathbb{Z}^d$ for $\Lambda = A\mathbb{Z}^d$, where A is a non-singular matrix describing Λ.

In 1958 I. M. Gelfand and A G. Kostyuchenko introduced Gelfand triples in their study of the spectral theory of self-adjoint operators [22]. They were motivated by the work of Dirac on the foundations of quantum mechanics and Schwartz's theory of distributions.

An important result of linear algebra is the theorem on the existence of eigenvectors for any self-adjoint linear operator A on $\mathbb{R}^d$. The situation changes drastically when one passes from the finite to the infinite-dimensional case, since it can happen that a unitary operators does not have any (non-zero) eigenvector. Particular examples of such operators are the translation operator T_x and the modulation operator M_ω on $L^2(\mathbb{R}^d)$.

Let us present an easy argument showing that the translation operator $T_x, x \neq 0$, has no eigenvectors in $L^2(\mathbb{R}^d)$. Assume that $f \in L^2(\mathbb{R}^d)$ satisfies

$$T_x f(t) = a f(t), \tag{30}$$

which by, the Fourier transform, is equivalent to

$$M_{-x}\hat{f}(\omega) = a\hat{f}(\omega) \quad \text{a.e..} \tag{31}$$

But this is only possible if the function $\hat{f}$ equals zero a.e., up to the points with $e^{2\pi i\omega x} \neq a$, i.e., it differs from zero only on a set of measure zero, hence $\hat{f} = 0$, and finally $f = 0 \in L^2(\mathbb{R}^d)$. In other words, the translation operator T_x does not have eigenvectors in the space $L^2(\mathbb{R}^d)$. On the other hand, we are not too far off with the claim that T_x has the eigenvectors $e^{-2\pi it\omega}$ corresponding to the eigenvalue $e^{2\pi ix\omega}$, and the claim that any function f in $L^2(\mathbb{R}^d)$ can be (kind of) expanded in terms of the eigenvectors $e^{-2\pi it\omega}$, by suitable interpretation of the inversion formula for the Fourier transform (valid pointwise for $f \in S_0(\mathbb{R}^d)$):

$$f(t) = \int_{\mathbb{R}^d} \hat{f}(\omega)e^{2\pi it\omega}d\omega. \tag{32}$$

Furthermore, the action of the translation operator is given by

$$T_x f(t) = \int_{\mathbb{R}^d} e^{2\pi ix\omega}\hat{f}(\omega)e^{2\pi it\omega}d\omega,$$

which is a continuous analog of the spectral decomposition of a self-adjoint operator in $\mathbb{R}^d$.

More concretely, the system of eigenfunctions $\{e^{-2\pi it\omega} : \omega \in \widehat{\mathbb{R}^d}\}$ is complete in the sense that for any function f in $L^2(\mathbb{R}^d)$ Parseval's equality holds

$$\int_{\mathbb{R}^d} |f(t)|^2 dt = \int_{\mathbb{R}^d} |\hat{f}(\omega)|^2 d\omega.$$

The obvious problem is the fact that $L^2(\mathbb{R}^d)$ does not contain the system of eigenvectors of the translation operator T_x. But they can be considered as linear functionals on $S_0(\mathbb{R}^d)$. This as well es several similar observations suggests to study operators on a Hilbert space via a dense subspace and its associated dual space. In our example it is actually possible to start from $S_0(\mathbb{R}^d)$ and construct $L^2(\mathbb{R}^d)$ as completion of $S_0(\mathbb{R}^d)$ with respect to norm corresponding to the usual scalar product $\langle f, g\rangle = \int_{\mathbb{R}^d} f(t)\overline{g(t)}dt$.

In this context it turns out that $S_0(\mathbb{R}^d)$ has the important additional property that both δ-distributions and the pure frequencies $\chi_\omega(x) =$

$e^{-2\pi i x\omega}$ (for all $\omega \in \mathbb{R}^d$) are in a natural way elements of $S_0'(\mathbb{R}^d)$, i.e. define bounded linear functionals on $S_0(\mathbb{R}^d)$. As a consequence we are now in a situation similar to the one inspiring Gelfand to introduce what is nowadays called a Gelfand triple. The main idea is the observation, that a triple of spaces – consisting of the Hilbert space itself, a small (topological vector) space contained in the Hilbert space, and its dual – allows a much better description of the situation. The advantage in our case is the fact that we can even consider a Banach space, namely $S_0(\mathbb{R}^d)$. Hence we can work with the following formal definition:

Definition 15: A (Banach) *Gelfand triple* consists of some Banach space $(B, \|.\|_B)$, which is continuously and densely embedded into some Hilbert space $\mathcal{H}$, which in turn is w^*-continuously and densely embedded into the dual Banach space $(B', \|.\|')$.

We shall use the symbol $(B, \mathcal{H}, B')$ for such a triple of spaces. In this setting the inner product on $\mathcal{H}$ extends in a natural way to a pairing between B and B' (producing an anti-linear functional of the same norm).

As another consequence we mention an extension of an eigenvector of a bounded operator on a Hilbert space $\mathcal{H}$. Let A be a linear operator on a Banach space B. Then a linear functional F is a *generalized eigenvector* of A to the eigenvalue λ if

$$F(Af) = \lambda F(f), \qquad \text{for all } \; f \in B.$$

This notion allows to interpret the characters $\chi_\omega(x) = e^{-2\pi i \omega x}$ as generalized eigenvectors for the translation operator T_x on $S_0(\mathbb{R}^d)$. Furthermore the set of generalized eigenvectors $\{\chi_\omega : \omega \in \mathbb{R}^d\}$ is complete by Plancherel's theorem, i.e., $\hat{f}(\omega) = \langle \chi_\omega, f \rangle = 0$ for all $\omega \in \mathbb{R}^d$, this implies $f \equiv 0$. This suggests to think of the Fourier transform of f at frequency ω as the evaluation of the linear functional $\langle \chi_\omega, f \rangle$.

The treatment of the translation operator T_x on $L^2(\mathbb{R}^d)$ is a particular case of a general theorem by Gelfand, that for any self-adjoint operator A on a Hilbert space $\mathcal{H}$ there exists a nuclear space and a complete system of generalized eigenvectors, see [23]. The advantage of the approach presented here is that instead of a (maybe complicated) nuclear topological vector space, a relatively simple-minded Banach space can be used.

The introduction of Gelfand triples does not only offer a better description of a self-adjoint operator but it allows also simplification of proofs. For example, in the discussion of the Fourier transform $\mathcal{F}$, the latter is considered as an object on $S_0(\mathbb{R}^d)$ where everything is well-defined, and Parseval's formula and taking the inverse Fourier transform are justified by the nice properties of $S_0(\mathbb{R}^d)$. By a density argument we get all properties of the Fourier transform on the level of $L^2(\mathbb{R}^d)$. And we obtain an extension of the Fourier transform to $S_0'(\mathbb{R}^d)$, the so-called *generalized Fourier transform*, by duality.

The preceding discussion suggests the following lemma which says that assertions for an operator on the S_0-level are actually statements for $L^2(\mathbb{R}^d)$ and S_0', respectively.

Lemma 16: *The Fourier transform $\mathcal{F}$ on $\mathbb{R}^d$ has the following properties:*

(1) *$\mathcal{F}$ is an isomorphism from $S_0(\mathbb{R}^d)$ to $S_0(\widehat{\mathbb{R}^d})$,*
(2) *$\mathcal{F}$ is a unitary map between $L^2(\mathbb{R}^d)$ and $L^2(\widehat{\mathbb{R}^d})$,*
(3) *$\mathcal{F}$ is a weak* (as well as a norm-to-norm) continuous bijection from $S_0'(\mathbb{R}^d)$ to $S_0'(\widehat{\mathbb{R}^d})$.*

Furthermore we have that Parseval's formula

$$\langle f, g \rangle = \langle \hat{f}, \hat{g} \rangle \tag{33}$$

is valid for $(f, g) \in S_0(\mathbb{R}^d) \times S_0'(\mathbb{R}^d)$, and therefore on each level of the Gelfand triple $(S_0, L^2, S_0')(\mathbb{R}^d)$.

The properties of Fourier transform are expressed by the *Gelfand bracket*

$$\langle f, g \rangle_{(S_0, L^2, S_0')(\mathbb{R}^d)} = \langle \hat{f}, \hat{g} \rangle_{(S_0, L^2, S_0')(\widehat{\mathbb{R}}^d)} \tag{34}$$

which combines the functional brackets of Banach spaces and of the inner-product for the Hilbert space.

The Fourier transform is a prototype for the notion of a Gelfand triple isomorphism.

Definition 17: If $(B_1, \mathcal{H}_1, B_1')$ and $(B_2, \mathcal{H}_2, B_2')$ are Gelfand triples then an operator A is called a *[unitary] Gelfand triple isomorphism* if

(1) A is an isomorphism between B_1 and B_2.
(2) A is a [unitary operator resp.] isomorphism from $\mathcal{H}_1$ to $\mathcal{H}_2$.

(3) A extends to a weak* isomorphism as well as a norm-to-norm continuous isomorphism between B_1' and B_2'.

In this terminology the Fourier transform is a unitary Gelfand triple isomorphism (actually an automorphism) on the Gelfand triple $(S_0, L^2, S_0')(\mathbb{R}^d)$. In the following lemma we give conditions for the extension of a linear mapping given on $S_0(\mathbb{R}^d)$ to a unitary mapping on $L^2(\mathbb{R}^d)$.

Lemma 18: *(cf. [19]) Let U be a unitary mapping from $L^2(\mathbb{R}^d)$ to $L^2(\mathbb{R}^d)$. The mapping U extends to a Gelfand triple isomorphism between $(S_0, L^2, S_0')(\mathbb{R}^d)$ and $(S_0, L^2, S_0')(\mathbb{R}^d)$ if and only if the restriction of U to $S_0(\mathbb{R}^d)$ defines a bounded bijective linear mapping from $S_0(\mathbb{R}^d)$ onto itself.*

Due to this lemma we only have to check the properties of U at the S_0-level, i.e., to verify the existence of some $C > 0$ such that

$$\|Uf\|_{S_0(\mathbb{R}^d)} \leq C\|f\|_{S_0(\mathbb{R}^d)}. \tag{35}$$

The discussion of the Fourier transform $\mathcal{F}$ on the Gelfand triple $(S_0, L^2, S_0')(\mathbb{R}^d)$ allows to think of $\mathcal{F}$ as a bounded operator between $S_0(\mathbb{R}^d)$ and $S_0'(\mathbb{R}^d)$ with a distributional kernel $k(t, \omega) = e^{-2\pi i t\omega}$. The existence of a distributional kernel for any bounded operator between $S_0(\mathbb{R}^d)$ and $S_0'(\mathbb{R}^d)$ is *kernel theorem* for $S_0(\mathbb{R}^d)$ (cf. [16], Thm. 7.4.2). Before we give a precise description of this important fact, we recall the notion of a Wilson basis. With the help of a Wilson basis we can adapt a linear algebra reasoning to the infinite-dimensional setting.

In 1991 Daubechies, Jaffard, and Journé [8] followed an idea of Wilson in their construction of an orthonormal basis from a Gabor system $\mathcal{G}(g, \Lambda)$ of $L^2(\mathbb{R}^d)$. Wilson suggested that the building blocks $\pi(x, \omega)g$ of an orthonormal basis of $L^2(\mathbb{R}^d)$ should be symmetric in ω and should be concentrated at ω and $-\omega$.

Definition 19: For $g \in L^2$ the associated Wilson system $\mathcal{W}(g)$ consists of the functions

$$\psi_{k,n} = c_n T_{\frac{k}{2}}(M_n + (-1)^{k+n} M_{-n})g, \quad (k, n) \in \mathbb{Z} \times \mathbb{N}_0,$$

where $c_0 = \frac{1}{2}$ and $c_n = \frac{1}{\sqrt{2}}$ for $n \geq 1$, $\psi_{k,0} = T_k g$ and $\psi_{2k+1,0} = 0$ for $k \in \mathbb{Z}$.

They proved the following theorem which shows a method for the construction of a Wilson basis from a Gabor system $\mathcal{G}(g, \frac{1}{2}\mathbb{Z} \times \mathbb{Z})$. Later Feichtinger, Gröchenig, and Walnut [14] showed that Wilson systems provide an unconditional basis for $S_0(\mathbb{R}^d)$ and $S_0'(\mathbb{R}^d)$ endowed with the w^*-topology (actually, for all modulation spaces $M_m^{p,q}(\mathbb{R}^d)$ with $p, q < \infty$). Therefore Wilson systems provide us with a natural class of bases for time-frequency analysis. The existence of an unconditional basis for $S_0(\mathbb{R}^d)$ will be very helpful in our discussion of the kernel theorem for $S_0(\mathbb{R}^d)$ and its construction relies heavily on the functorial properties of S_0, cf. [16].

Theorem 20: *Let $\mathcal{G}(g, \frac{1}{2}\mathbb{Z} \times \mathbb{Z})$ be a tight frame for $L^2(\mathbb{R})$ with $\|g\| = 1$ and $g(x) = \overline{g(-x)}$. Then the Wilson system $\mathcal{W}(g)$ is an orthonormal basis of $L^2(\mathbb{R})$.*

As a corollary we get Wilson bases for $L^2(\mathbb{R}^d)$ by taking tensor products.

Corollary 21: *Let $\mathcal{W}(g)$ be a Wilson basis for $L^2(\mathbb{R})$ and define $\mathbf{\Psi}_{k,n} = \prod_{j=1}^{d} \psi_{r_j,s_j}$ for $(r,s) \in \mathbb{Z}^d \times \mathbb{N}_0^d$. Then $\mathbf{\Psi}_{k,n}$ is an orthonormal basis for $L^2(\mathbb{R}^d)$.*

In applications of mathematics one often has to deal with linear systems. In the discrete and finite case each linear system is a linear mapping from the input space $\mathbb{R}^n$ into the output space $\mathbb{R}^m$ of our system and its action is given by matrix multiplication after a choice of bases in $\mathbb{R}^n$ and $\mathbb{R}^m$, respectively (similarly from $\mathbb{C}^n$ to $\mathbb{C}^m$ using complex matrices).

A linear system in infinite dimensions may be considered as a continuous analog of matrix multiplication (replacing summation by integration), i.e.,

$$g(x) = Kf(x) = \int_{\mathbb{R}^d} k(x,y) f(y) dy.$$

We can think of the input values $f(y)$ as being listed in an infinite column vector and $k(x,y)$ as an infinite matrix, the so-called *kernel* of K, and the integral $\int_{\mathbb{R}^d} k(x,y) f(y) dy$ providing the entries of the output vector in the expected way. In signal processing, such a model is known as a *linear time-variant system.*

For a wide range of function spaces (covering practically all cases relevant for applications) and by means of the use of generalized functions, this analogy can be given a precise mathematical meaning. The natural way of describing this context is via so-called *kernel theorems*. Although only Hilbert Schmidt operators can be described as integral operators with L^2-kernels, every bounded linear system A on $L^2(\mathbb{R}^d)$ can be uniquely described by some distributional kernel $K \in S_0'(\mathbb{R}^{2d})$.

The notion of Gelfand triples suggests to consider bounded linear operators between arbitrary L^p and L^q-spaces (with $p < \infty$) as bounded operators from $S_0(\mathbb{R}^d)$ to $S_0'(\mathbb{R}^d)$ (by trivial restriction of the range). Since Kf is only a distribution, we can describe it only indirectly by applying the output distribution to some test function $g \in S_0(\mathbb{R}^d)$.

Suppose we have an integral operator K with distributional kernel k on $S_0(\mathbb{R}^d)$, i.e., we think of K in a weak sense

$$\langle Kf, g\rangle = \langle k, g \otimes f\rangle, \quad f, g \in S_0(\mathbb{R}^d),$$

where $(g \otimes f)(x, y)$ denotes the tensor product $g(x)f(y)$, then K is a bounded operator between $S_0(\mathbb{R}^d)$ and $S_0'(\mathbb{R}^d)$. Since by duality we deduce that

$$|\langle Kf, g\rangle| = |\langle k, f \otimes g\rangle| \leq \|k\|_{S_0'} \|f \otimes g\|_{S_0} = \|k\|_{S_0'} \|f\|_{S_0} \|g\|_{S_0}$$

is true for all $g \in S_0(\mathbb{R}^d)$, we have that $Kf \in S_0'(\mathbb{R}^d)$. Therefore the operator K is bounded between $S_0(\mathbb{R}^d)$ and $S_0'(\mathbb{R}^d)$ with the following estimate for the operator norm of K:

$$\|K\|_{op} \leq \|k\|_{S_0'} .$$

The non-trivial aspect of the kernel theorem is that the converse is true.

Theorem 22: *If K is a bounded operator from $S_0(\mathbb{R}^d)$ to $S_0'(\mathbb{R}^d)$, then there exists a unique kernel $k \in S_0'(\mathbb{R}^{2d})$ such that $\langle Kf, g\rangle = \langle k, g \otimes f\rangle$ for $f, g \in S_0(\mathbb{R}^d)$.*

We only sketch a proof and refer the interested reader to the book of Gröchenig [26] for the technical details.

We define the infinite matrix $\mathbf{a} = \left(a_{(l,m),(r,s)}\right)$ of the operator K with respect to a multivariate Wilson basis $\mathcal{W}(g)$ by

$$a_{(l,m),(r,s)} = \langle K\mathbf{\Psi}_{r,s}, \mathbf{\Psi}_{l,m}\rangle. \tag{36}$$

Then the matrix $\mathbf{a}$ is bounded from $\ell^1(\mathbb{Z}^d \times \mathbb{N}_0^d)$ to $\ell^\infty(\mathbb{Z}^d \times \mathbb{N}_0^d)$. Therefore, we can define a kernel k for K as in linear algebra, by

$$k = \sum_{l,m,r,s} a_{(l,m),(r,s)} \mathbf{\Psi}_{l,m} \otimes \mathbf{\Psi}_{r,s}. \tag{37}$$

Now, we know that $\{\mathbf{\Psi}_{l,m} \otimes \mathbf{\Psi}_{r,s}\}$ is an orthonormal basis for $L^2(\mathbb{R}^{2d})$ which yields that $k \in S_0'(\mathbb{R}^{2d})$ with weak*-convergence of the sum.

An important corollary of the preceding discussion is the following observation.

Corollary 23: *Let $(\mathbf{\Psi}_{\mathbf{k},\mathbf{n}})$ be an orthonormal Wilson basis for $L^2(\mathbb{R}^d)$. Then the coefficient mapping $D : f \mapsto \langle f, \mathbf{\Psi}_{k,n}\rangle$ induces a Gelfand triple isomorphism between $(S_0, L^2, S_0')(\mathbb{R}^d)$ and $(\ell^1, \ell^2, \ell^\infty)(\mathbb{Z}^d \times \mathbb{N}^d)$.*

Proof: Since $(\mathbf{\Psi}_{k,n})$ is an orthonormal basis of $L^2(\mathbb{R}^d)$ the analysis operator $f \mapsto \langle f, \mathbf{\Psi}_{k,n}\rangle$ is an isomorphism between $L^2(\mathbb{R}^d)$ and $\ell^2(\mathbb{Z}^d \times \mathbb{N}^d)$. The Wilson system $(\mathbf{\Psi}_{k,n})$ is an unconditional basis for $S_0(\mathbb{R}^d)$ and therefore the analysis operator gives an isomorphism between $S_0(\mathbb{R}^d)$ and $\ell^1(\mathbb{Z}^d \times \mathbb{N}^d)$. By duality we obtain that $S_0'(\mathbb{R}^d)$ is isomorphic to $\ell^\infty(\mathbb{Z}^d \times \mathbb{N}^d)$. □

8. The Spreading Function

The notion of a Gelfand triple has turned out to be a very fruitful concept for investigations in Gabor analysis, see [16], [7], [10]. In this section we present some results of Feichtinger and Kozek on Gelfand triples for time-frequency analysis. All these results have their origin in the search of a mathematical framework for problems in signal analysis. Many problems in applications are modeled as linear time-variant systems (LTV). In the last section we learned that a LTV is just an integral operator K acting on signals with finite energy,

$$Kf(x) = \int_{\mathbb{R}^d} k(x,y)f(y)dy, \qquad f \in L^2(\mathbb{R}^d). \tag{38}$$

The quality of an integral operator K on $L^2(\mathbb{R}^d)$ relies on properties of its kernel k. For example, integrability conditions on k yield classes of nice operators. The most prominent class of operators, the *Hilbert-Schmidt* operators $\mathcal{HS}$, are defined in terms of integrability conditions. Namely, an integral operator K on $L^2(\mathbb{R}^d)$ is a *Hilbert-Schmidt* operator if $k \in L^2(\mathbb{R}^d \times \mathbb{R}^d)$. From the analogy between integral operators and matrices we see that Hilbert-Schmidt operators are a generalization of the space of linear mappings on a finite dimensional vector space V with respect to the Frobenius norm, see (10) in Section 2.

The class of Hilbert-Schmidt operators $\mathcal{HS}$ has a natural inner product. Let $K_1, K_2 \in \mathcal{HS}$ with kernels k_1, k_2, respectively. Then

$$\langle K_1, K_2 \rangle_{\mathcal{HS}} := \langle k_1, k_2 \rangle_{L^2(\mathbb{R}^d \times \mathbb{R}^d)} \tag{39}$$

defines an inner product on $\mathcal{HS}$. The associated *Hilbert-Schmidt norm* $\|K\|_{\mathcal{HS}} := \big(\langle K, K \rangle_{\mathcal{HS}}\big)^{1/2}$ gives $\mathcal{HS}$ the structure of a Hilbert space [40]. Furthermore we recall that every Hilbert-Schmidt operator in $\mathcal{HS}$ is a compact operator on $L^2(\mathbb{R}^d)$. Recall that a compact operator K on $L^2(\mathbb{R}^d)$ is of Hilbert-Schmidt type if and only if there exists an orthonormal basis $(e_n)_{n\in\mathbb{N}}$ in $L^2(\mathbb{R}^d)$ and a sequence of scalars $(\lambda_n)_{n\in\mathbb{N}} \in \ell^2(\mathbb{N})$ such that

$$Kf = \sum_{n\in\mathbb{N}} \lambda_n \langle e_n, f \rangle e_n. \tag{40}$$

The sequence of scalars $(\lambda_n)_{n\in\mathbb{N}}$ are actually the eigenvalues of K and $\|K\|_{\mathcal{HS}} = \big(\sum_{n\in\mathbb{N}} |\lambda_n|^2\big)^{1/2}$. The space of Hilbert-Schmidt operators $\mathcal{HS}$ is not closed in the C^*-algebra $\mathcal{K}$ of compact operators on $L^2(\mathbb{R}^d)$ with respect to the operator norm, and there exist compact operators which are not of Hilbert-Schmidt type. But $\mathcal{HS}$ is a two-sided ideal in $\mathcal{K}$.

If we choose as orthonormal basis of $L^2(\mathbb{R}^d)$ a Wilson basis $(\boldsymbol{\Psi}_{\mathbf{k},\mathbf{n}})$, then the preceding observations lead to an isomorphism between $\mathcal{HS}$ and $\ell^2(\mathbb{Z}^d \times \mathbb{N}^d)$. Now we can make use of the concept of Gelfand triples, but this time we take the Hilbert-Schmidt operators as Hilbert space of an *Operator Gelfand triple*. We observe that the kernel theorem for $S_0(\mathbb{R}^d)$ provides us with another class of operators with "smooth kernels". We write $\mathcal{L}$ for the space of bounded linear operators on a Banach space B. One finds that $K \in \mathcal{L}(S_0'(\mathbb{R}^d), S_0(\mathbb{R}^d))$ can be identified with kernels $k \in S_0'(\mathbb{R}^{2d})$ which is dense in $\mathcal{HS}$. But the class of Hilbert-Schmidt operators $\mathcal{HS}$ is dense in $\mathcal{L}(S_0(\mathbb{R}^d), S_0'(\mathbb{R}^d))$ and therefore $(\mathcal{L}(S_0'(\mathbb{R}^d), S_0(\mathbb{R}^d)), \mathcal{HS}, \mathcal{L}(S_0(\mathbb{R}^d), S_0'(\mathbb{R}^d)))$ is indeed a Gelfand triple. In this setting the kernel theorem can be interpreted as a unitary Gelfand triple isomorphism between this triple of operators and their kernels in $(S_0, L^2, S_0')(\mathbb{R}^d \times \mathbb{R}^d)$. There is another Gelfand triple isomorphism that associates the $\mathcal{HS}$ Gelfand triple with the Gelfand triple $(S_0, L^2, S_0')(\mathbb{R}^d \times \widehat{\mathbb{R}}^d)$: the so-called *spreading symbol* of operators.

As a motivation we discuss a problem of great practical interest: communication with cellular phones. In modern communication cellular phones play a crucial rule in everyday life. How do engineers solve the problem of transmitting a signal f from a sender A to a receiver B? In the most general situation, sender A and receiver B move in different directions with certain velocities, which leads to a variation of the path lengths of the transmitted signal f and, due to the Doppler effect, to a change of frequencies. Therefore B receives a signal of the following form

$$\tilde{f} = \iint_{\mathbb{R}^2} \eta(K)(x,\omega) M_\omega T_x f \, dx d\omega, \tag{41}$$

where the function $\eta(K)$ models the effect of the channel by the amount of time-frequency shifts arising as just described, applied to the signal f.

The receiver B is not interested in the signal $\tilde{f}$ but in the original signal f. From a mathematical point of view, $\tilde{f}$ is just the action of an operator K on the signal f, i.e., $\tilde{f} = Kf$. In this picture B has to invert the operator K to get the information contained in the signal f. Operators of this form are called *pseudo-differential operators* and arise naturally in many problems of physics, engineering and mathematics. The function $\eta(K)$ is the so-called *spreading function* of the operator K. In the following, we look for conditions on the spreading function $\eta(K)$ which allow an inversion of our pseudo-differential operator K.

First the equation (41) suggests a decomposition of a general operator K on $L^2(\mathbb{R}^d)$ as a continuous superposition of time-frequency shifts.

$$K = \iint_{\mathbb{R}^{2d}} \eta(K)(x,\omega) M_\omega T_x \, dx d\omega. \tag{42}$$

We already know such a decomposition of the identity operator on $L^2(\mathbb{R}^d)$ since this is the inversion formula for the STFT:

$$I_{L^2(\mathbb{R}^d)} = \frac{1}{\langle g, \gamma \rangle} \iint_{\mathbb{R}^{2d}} V_g f(x,\omega) M_\omega T_x \, dx d\omega \tag{43}$$

for $g, \gamma \in L^2(\mathbb{R}^d)$ with $\langle g, \gamma \rangle \neq 0$.

The non-commutativity of translation and modulation operators on $L^2(\mathbb{R}^d)$ leads to a twisted convolution of the spreading functions of two operators K and L. Let $K, L \in \mathcal{L}(S_0, S_0')$ and $\eta(K), \eta(L)$ their spreading functions. Then the spreading function of the composition KL is given by *twisted convolution* of $\eta(K)$ and $\eta(L)$:

$$\eta(KL)(x,\omega) = \iint_{\mathbb{R}^2} \eta(K)(x',\omega')\eta(L)(x-x',\omega-\omega')e^{-2\pi i x'(\omega-\omega')}d\omega'. \quad (44)$$

The spreading function of the adjoint operator K^* is given by

$$\eta(K^*)(x,\omega) = \overline{\eta(K)(-x,-\omega)} \cdot e^{-2\pi i x\omega} \quad (45)$$

and therefore leads to a noncommutative involution. Later we will return to this topic in the context of Gröchenig/Leinert's resolution of the "irrational case"-conjecture [27].

The relation between the kernel k of an operator K from the Gelfand triple $(\mathcal{L}(S_0, S_0'), \mathcal{HS}, \mathcal{L}(S_0', S_0))$ and its spreading function $\eta(K)$ is given by

$$\eta(K)(x,\omega) = \int_{\mathbb{R}^d} k(y, y-x)e^{-2\pi i y\omega}dy, \quad (46)$$

which is very useful in the calculation of the spreading function of an operator K. It can be interpreted literally at the lowest level (integrals etc. exist), and extends by continuity to the "upper levels". Moreover, it can be described by the fact that it is the unique Gelfand triple isomorphism which maps TF-shift operators onto the corresponding Dirac measures in the TF-plane (hence reproducing exactly the situation we had in the finite case).

The spreading function of an operator K is an object living on the time-frequency plane $\mathbb{R}^d \times \widehat{\mathbb{R}}^d$. Therefore a further understanding of its properties is necessary according to the structure of $\mathbb{R}^d \times \widehat{\mathbb{R}}^d$ which is closely related to the structure of the Euclidean plane $\mathbb{R}^d \times \mathbb{R}^d$. Namely, the time-frequency plane is a symplectic manifold, i.e., there exists a non-degenerate 2-form $\Omega(X,Y) = y\cdot\omega - x\cdot\eta$ for two points $X = (x,\omega), Y = (y,\eta)$ in $\mathbb{R}^d \times \widehat{\mathbb{R}}^d$. Since Ω is non-degenerate there is a unique invertible skew-symmetric linear operator $\mathcal{J}$ on $\mathbb{R}^d \times \widehat{\mathbb{R}}^d$ such that the symplectic form Ω and the Euclidian inner product are related as follows: $\Omega(X,Y) = \langle \mathcal{J}X, Y\rangle$ for all $X, Y \in \mathbb{R}^d \times \widehat{\mathbb{R}}^d$. This implies an important fact about the characters of $\mathbb{R}^d \times \widehat{\mathbb{R}}^d$. Namely, the characters are given by $\{\chi_s(X,Y) = e^{2\pi i\Omega(X,Y)} | X \in \mathbb{R}^d \times \widehat{\mathbb{R}}^d\}$ for a fixed $Y \in \mathbb{R}^d \times \widehat{\mathbb{R}}^d$. Therefore it is natural to analyze a function F on $\mathbb{R}^d \times \widehat{\mathbb{R}}^d$ with the *symplectic Fourier Transform*

$$\mathcal{F}_s F(X) = \iint_{\mathbb{R}^d \times \widehat{\mathbb{R}}^d} F(Y)e^{2\pi i\Omega(X,Y)}dY \quad (47)$$

instead of the Fourier transform $\mathcal{F}$ induced by the standard inner-product $\langle\cdot,\cdot\rangle$ on $\mathbb{R}^d \times \mathbb{R}^d$. From the relation between symplectic form and inner-product we obtain that the symplectic Fourier transform $\mathcal{F}_s$ is just a Fourier transform followed by a rotation by $\frac{\pi}{2}$ since $\mathcal{J}$ describes a rotation by $\frac{\pi}{2}$ around the origin of $\mathbb{R}^d \times \mathbb{R}^d$. This fact allows us to derive similar statements for the symplectic Fourier transform as for the Euclidian Fourier transform.

(1) $\mathcal{F}_s$ is a unitary mapping from $L^2(\mathbb{R}^d \times \widehat{\mathbb{R}}^d)$ onto $L^2(\mathbb{R}^d \times \widehat{\mathbb{R}}^d)$.
(2) $\mathcal{F}_s^{-1} = \mathcal{F}_s$ (involutive property).
(3) $\mathcal{F}_s\left(S_0(\mathbb{R}^d \times \widehat{\mathbb{R}}^d)\right) = S_0(\mathbb{R}^d \times \widehat{\mathbb{R}}^d)$.

By duality we obtain that

Proposition 24: *The symplectic Fourier transform $\mathcal{F}_s$ defines a unitary Gelfand triple automorphism on $(S_0, L^2, S_0')(\mathbb{R}^d \times \widehat{\mathbb{R}}^d)$.*

Another reason for our choice of $S_0(\mathbb{R}^{2d})$ as space of test functions is that the Poisson summation formula for symplectic Fourier transform holds pointwise and with absolute convergence. Recently, we have shown that the Fundamental Identity of Gabor Analysis can be derived by an application of Poisson summation to a product of two STFT's:

Theorem 25: *Let Λ a lattice in $\mathbb{R}^d \times \widehat{\mathbb{R}}^d$ with adjoint lattice Λ° and $F \in S_0(\mathbb{R}^{2d})$. Then*

$$\sum_{\lambda\in\Lambda} F(\lambda) = \frac{1}{|\Lambda|} \sum_{\lambda^\circ\in\Lambda^\circ} \mathcal{F}_s F(\lambda^\circ) \tag{48}$$

holds pointwise and with absolute convergence on both sides.

The spreading function is an important tool for the description of (slowly) time-variant channels in communication theory, but it is not the only symbol associated with a linear operator. In the theory of pseudo-differential operators the *Kohn-Nirenberg symbol* (KN), denoted by $\sigma(K)$, is used for an operator $K \in (S_0, L^2, S_0')(\mathbb{R}^d)$. It is defined as the symplectic Fourier transform of the spreading function $\eta(K)$:

$$\sigma(x,\omega) = \mathcal{F}_s\eta(K) = \iint_{\mathbb{R}^d\times\widehat{\mathbb{R}}^d} \eta(K)e^{2\pi i(y\cdot\omega - x\cdot\eta)}dyd\eta, \quad (x,\omega)\in\mathbb{R}^d\times\widehat{\mathbb{R}}^d. \tag{49}$$

If $Kf(x) = \int_{\mathbb{R}^d} k(x,y)f(y)dy$ then $\sigma(K) = \int_{\mathbb{R}^d} k(x, x-y)e^{-2\pi i y\cdot\omega}dy$. In signal analysis $\sigma(K)$ was introduced by Zadeh and is called the *time-varying transfer function* of a system modeled by K. As an example we mention the KN symbol of a rank-one operator $f \otimes \overline{g}$, which describes the mapping $h \mapsto \langle h, g\rangle f$, is equal to

$$\sigma(f \otimes \overline{g})(x,\omega) = f(x)\overline{\hat{g}(\omega)}e^{-2\pi i x\cdot\omega}, \qquad (x,\omega) \in \mathbb{R}^d \times \widehat{\mathbb{R}}^d, \tag{50}$$

the Rihaczek distribution of f against g. For $f, g \in S_0(\mathbb{R}^d)$ we have that the KN-symbol $\sigma(f \otimes \overline{g})$ is in $S_0(\mathbb{R}^d \times \widehat{\mathbb{R}}^d)$ which in turn implies (using the last equation) that $(x,\omega) \mapsto e^{2\pi i x\cdot\omega}$ is a pointwise multiplier on $S_0(\mathbb{R}^d \times \widehat{\mathbb{R}}^d)$.

After these preparations we can state one of our main results:

Theorem 26: *The spreading function $K \mapsto \eta(K)$ is a unitary Gelfand triple isomorphism from $(\mathcal{L}(S_0, S_0'), \mathcal{HS}, \mathcal{L}(S_0', S_0))$ to $(S_0, L^2, S_0')(\mathbb{R}^d \times \widehat{\mathbb{R}}^d)$.*

Corollary 27: *The KN symbol of K induces a unitary Gelfand triple isomorphism between $(\mathcal{L}(S_0, S_0'), \mathcal{HS}, \mathcal{L}(S_0', S_0))$ and $(S_0, L^2, S_0')(\mathbb{R}^d \times \widehat{\mathbb{R}}^d)$.*

Another consequence of the preceding theorem are the following Gelfand-bracket identities for $K_1, K_2 \in (\mathcal{L}(S_0, S_0'), \mathcal{HS}, \mathcal{L}(S_0', S_0))$:

$$\langle K_1, K_2\rangle_{(\mathcal{B},\mathcal{HS},\mathcal{B}')} = \langle \eta(k_1), \eta(k_2)\rangle_{(S_0,L^2,S_0')(\mathbb{R}^d\times\widehat{\mathbb{R}}^d)} \tag{51}$$

$$= \langle \sigma(k_1), \sigma(k_2)\rangle_{(S_0,L^2,S_0')(\mathbb{R}^d\times\widehat{\mathbb{R}}^d)}, \tag{52}$$

with $\mathcal{B} = \mathcal{L}(S_0, S_0')$ and $\mathcal{B}' = \mathcal{L}(S_0', S_0)$ respectively.

The KN symbol of a rank-one operator $f \otimes \overline{g}$, is the Rihaczek distribution and by an application of the (inverse) symplectic Fourier transform we get another time-frequency distribution: the STFT!

Lemma 28: *For $f, g \in S_0(\mathbb{R}^d)$ the rank-one operator $f \otimes \overline{g}$ has a kernel in $S_0(\mathbb{R}^d)$. Moreover the corresponding spreading function is*

$$\eta(f \otimes \overline{g})(x,\omega) = \int_{\mathbb{R}^d} f(x)\overline{g(y-x)}e^{-2\pi i y\cdot\omega}dy \tag{53}$$

and hence coincides with $V_g f \in S_0(\mathbb{R}^d \times \widehat{\mathbb{R}}^d)$.

In the light of this result the inversion formula for the STFT is a superposition of time-frequency shifts with the spreading function of the rank-one operator $g \otimes \overline{\gamma}$ for $g, \gamma \in L^2(\mathbb{R}^d)$ with $\langle g, \gamma \rangle \neq 0$:

$$f = \frac{1}{\langle g, \gamma \rangle} \iint_{\mathbb{R}^d \times \widehat{\mathbb{R}}^d} \eta(f \otimes \overline{g})(x, \omega) T_x M_\omega \gamma \, dx d\omega. \tag{54}$$

Recall that in analogy with the characters $\{\chi_\omega : \omega \in \widehat{\mathbb{R}^d}\}$, the time-frequency shifts $\{\pi(X) : X = (x, \omega) \in \mathbb{R}^d \times \widehat{\mathbb{R}}^d\}$ would be an orthonormal set with respect to the Hilbert-Schmidt inner product $\langle \cdot, \cdot \rangle_{\mathcal{HS}}$ and $\eta(f \otimes \overline{g})(x, \omega) = \langle f \otimes \overline{g}, \pi(x, \omega) \rangle_{\mathcal{HS}}$ but as in the case of Fourier transform, the building blocks $\pi(X)$ for $X \in \mathbb{R}^d \times \widehat{\mathbb{R}}^d$ of our orthonormal system $\{\pi(X) : X = (x, \omega) \in \mathbb{R}^d \times \widehat{\mathbb{R}}^d\}$ are not Hilbert-Schmidt. As in our treatment of the Fourier transform, it is not so important that the building blocks are elements of our Hilbert space but that they allow us to get expressions as they were an orthonormal set of elements in our Hilbert space.

As a first example we state a generalization of the inversion formula for the STFT from $L^2(\mathbb{R}^d)$ to the Gelfand triple $(S_0, L^2, S_0')(\mathbb{R}^d)$, where for $f \in S_0'(\mathbb{R}^d)$ the formula is interpreted in a weak sense.

Proposition 29: *Let $g, \gamma \in S_0(\mathbb{R}^d)$ with $\langle g, \gamma \rangle \neq 0$. Then*

$$f = \frac{1}{\langle g, \gamma \rangle} \iint_{\mathbb{R}^d \times \widehat{\mathbb{R}}^d} \eta(f \otimes \overline{g})(x, \omega) T_x M_\omega \gamma \, dx d\omega \tag{55}$$

holds for $f \in (S_0, L^2, S_0')(\mathbb{R}^d)$.

That is a special case of a general statement about the spreading function.

Theorem 30: *Any $K \in (\mathcal{L}(S_0, S_0'), \mathcal{HS}, \mathcal{L}(S_0', S_0))$ has a representation*

$$K = \iint_{\mathbb{R}^d \times \widehat{\mathbb{R}}^d} \langle K, \pi(x, \omega) \rangle_{\mathcal{L}(S_0, S_0')} \, \pi(x, \omega) \, dx d\omega \tag{56}$$

convergent in the strong resp. weak-sense. The (complex-valued) amplitude function arising in this context, i.e. $\eta(K)(x, \omega) = \langle K, \pi(x, \omega) \rangle_{\mathcal{L}(S_0, S_0')}$, is called the spreading distribution of the operator K.*

The basic tool in the proof is the fact that the spreading representation maps a time-frequency shift $\pi(X)$, for $X = (x, \omega) \in \mathbb{R}^d \times \widehat{\mathbb{R}}^d$, on the Dirac measure δ_X, i.e., $\eta(\pi(X)) = \delta_X$, and the relation between the spreading function and the kernel of an operator K.

The preceding theorem is the mathematical justification of a widely used statement that the spreading function of an operator K is a measure for the time-frequency content of K.

In our intuition we move an operator K over $\mathbb{R}^d \times \widehat{\mathbb{R}}^d$ and expect a simple relation between the original symbol of K and the symbol after a movement to $(x, \omega) \in \mathbb{R}^d \times \widehat{\mathbb{R}}^d$. The KN-symbol of an operator K is shifted by $T_{x,\omega}$ in the time-frequency plane.

Lemma 31: *Let K belong to one of the spaces $(\mathcal{L}(S_0', S_0), \mathcal{HS}, \mathcal{L}(S_0', S_0))$, then $\pi(x,\omega)K\pi(x,\omega)^*$, the conjugation of K by $\pi(x,\omega), (x,\omega) \in \mathbb{R}^d \times \widehat{\mathbb{R}}^d$, corresponds to translation of the KN symbol $\sigma(K)$,*

$$\sigma(\pi(x,\omega)K\pi(x,\omega)^*) = T_{(x,\omega)}(\sigma(K)). \tag{57}$$

This property of the KN symbol is of central importance in our study of the Gabor frame operator to which we devote the final part of this section. Let $\mathcal{G} = (g, \Lambda)$ be a Gabor system for a lattice $\Lambda \in \mathbb{R}^d \times \widehat{\mathbb{R}}^d$. Then the Gabor frame operator $S_{g,\Lambda}$ commutes with all time-frequency shifts of the lattice Λ, i.e.

$$\pi(\lambda)S_{g,\Lambda}\pi(\lambda)^* = S_{g,\Lambda}, \qquad \text{for all} \quad \lambda \in \Lambda. \tag{58}$$

This fact was the motivation for Feichtinger and Kozek to introduce the class of Λ-invariant operators [16].

Definition 32: Let Λ be a lattice in $\mathbb{R}^d \times \widehat{\mathbb{R}}^d$ and K an operator in $\mathcal{B}(\Lambda)$. Then K is called Λ-*invariant* if $\pi(\lambda)K = K\pi(\lambda)$ for all $\lambda \in \Lambda$.

In the following we want to find the support of the spreading function $\eta(K)$ of an Λ-invariant operator $K \in (\mathcal{L}(S_0, S_0'), \mathcal{HS}, \mathcal{L}(S_0', S_0))$. As a first step towards this result we study spreading representations of K on $\mathbb{R}^d \times \widehat{\mathbb{R}}^d$.

Lemma 33: *Let $K_1, K_2 \in \mathcal{L}(S_0, S_0')$ with spreading function $\eta(K_1), \eta(K_2)$, respectively. Then*

(1) $\eta(K_1K_2)(\lambda) = \iint_{\mathbb{R}^d \times \widehat{\mathbb{R}}^d} \eta(K_1)(\mu)\eta(K_2)(\lambda - \mu)\rho(\lambda - \mu, \mu)d\mu$ *with* $\rho(X,Y) = e^{2\pi i(y\cdot\omega - x\cdot\eta)}$ *for* $X = (x,\omega), Y = (y,\eta) \in \mathbb{R}^d \times \widehat{\mathbb{R}}^d$.

(2) *$supp(\eta(K_1)\eta(K_2)) \subset supp(K_1) + supp(K_2)$.*

(3) $|\eta(K_1K_2)| = |\eta(K_1)| * |\eta(K_2)|$ *for* $\eta(K_1), \eta(K_2) \in L^1_{loc}(\mathbb{R}^d \times \widehat{\mathbb{R}}^d)$.

The proof of (i) is a consequence of the commutation relation for time-frequency shifts and the fact that for $K_1 \in \mathcal{L}(S_0, S_0')$ and $K_2 \in \mathcal{L}(S_0, S_0')$ also $K_1 K_2 \in \mathcal{L}(S_0', S_0)$. Now each operator K in $\mathcal{L}(S_0, S_0')$ has an absolutely convergent spreading representation and therefore our result holds pointwise. The support condition follows from the analogous result for the ordinary convolution.

By abstract reasons, each Λ-invariant operator K has a representation in the set of all operators concentrated on $\Lambda^\circ = \{\lambda^\circ \in \mathbb{R}^d \times \widehat{\mathbb{R}}^d | \pi(\lambda)\pi(\lambda^\circ) = \pi(\lambda^\circ)\pi(\lambda)\}$ since K lies in the commutant of the (C^*, von Neumann) algebra generated by $\{\pi(\lambda) : \lambda \in \Lambda\}$. The set Λ° is the so-called *adjoint lattice*, since it is the annihilator subgroup of Λ for the symplectic Fourier transform $\mathcal{F}_s$, and if $\Lambda^\perp$ is the annihilator subgroup of Λ with respect to $\mathcal{F}$, then $\Lambda^\circ = \mathcal{J}\Lambda^\perp$.

The time-frequency invariance of $S_0(\mathbb{R}^d)$ implies that K and $\pi(\lambda)K$ are in the Gelfand triple $(\mathcal{L}(S_0, S_0'), \mathcal{HS}, \mathcal{L}(S_0', S_0))$ too. Therefore, the Λ-invariance of T translates into a periodicity condition for the symbol $\sigma(K)$

$$\sigma(K) = T_\lambda(\sigma(K)), \quad \lambda \in \Lambda\,. \tag{59}$$

This periodicity condition corresponds to a support condition for the spreading function since $\{e^{-2\pi\Omega(\lambda,\mu)} | \lambda \in \Lambda\}$ for a fixed $\mu \in \mathbb{R}^d \times \widehat{\mathbb{R}}^d$ is a group of characters on $\mathbb{R}^d \times \widehat{\mathbb{R}}^d$ yields that

$$\operatorname{supp}(\eta(K)) \subset \mathcal{J}\Lambda^\perp = \Lambda^\circ. \tag{60}$$

The fact that distributions in $S_0'(\mathbb{R}^d)$ with support in a discrete subgroup is a sum of Dirac measures with a bounded sequence of coefficients implies, that for some bounded sequence (c_{λ°) over Λ°

$$\eta(K) = \sum_{\lambda^\circ \in \Lambda^\circ} c_{\lambda^\circ} \delta_{\lambda^\circ} \tag{61}$$

with $c_{\lambda^\circ} = (K)_{\lambda^\circ} = \iint_{\mathbb{R}^d \times \widehat{\mathbb{R}}^d / \Lambda^\circ} \sigma(K)(\mu) e^{2\pi i \Omega(\lambda,\mu)} d\mu$.

Returning to the description in the operator domain we arrive at the following characterization.

Theorem 34: *Let $K \in (\mathcal{L}(S_0, S_0'), \mathcal{HS}, \mathcal{L}(S_0', S_0))$ and $\sigma(K)$ the KN symbol. Then $\sigma(K)$ is a Λ-periodic distribution with a symplectic Fourier transform supported on Λ°. Furthermore*

$$K = \sum_{\lambda^\circ \in \Lambda^\circ} (K)_{\lambda^\circ} \pi(\lambda^\circ). \tag{62}$$

Corollary 35: *The mapping $\sigma(K) \mapsto (K)_{\lambda^\circ}$ is a unitary Gelfand triple isomorphism between $(S_0, L^2, S_0')(\mathbb{R}^d \times \widehat{\mathbb{R}}^d/\Lambda)$ and $(\ell^1, \ell^2, \ell^\infty)(\Lambda^\circ)$.*

Note that the time-frequency invariance of $S_0(\mathbb{R}^d)$ implies the boundedness of K on $S_0(\mathbb{R}^d)$ since

$$\|K\|_{\mathcal{L}(S_0)} = \| \sum_{\lambda^\circ \in \Lambda^\circ} (K)_{\lambda^\circ} \pi(\lambda^\circ)\|_{\mathcal{L}(S_0)} \leq \sum_{\lambda^\circ \in \Lambda^\circ} |(K)_{\lambda^\circ}|. \tag{63}$$

The next theorem shows that for any Λ-invariant operator K with $\sigma(K) \in S_0'((\mathbb{R}^d \times \widehat{\mathbb{R}}^d)/\Lambda)$ there exists a prototype operator $P \in \mathcal{L}(S_0, S_0')$ such that periodization of P in the time-frequency plane corresponds to sampling of the spreading function $\eta(P)$ on Λ°.

Theorem 36: *Let K be a Λ-invariant operator with $\sigma(K) \in S_0'((\mathbb{R}^d \times \widehat{\mathbb{R}}^d)/\Lambda)$. Then there exists some $P \in \mathcal{L}(S_0, S_0')$ such that its periodization is exactly K*

$$K = \sum_{\lambda \in \Lambda} \pi(\lambda)\, P\, \pi(\lambda)^* = \frac{1}{|\Lambda|} \sum_{\lambda^\circ \in \Lambda^\circ} \langle P, \pi(\lambda^\circ)\rangle_{\mathcal{L}(S_0, S_0')}\ \pi(\lambda^\circ). \tag{64}$$

Remark 37: The preceding result is a discrete analog of our spreading representation for operators in $\mathcal{L}(S_0, S_0')$ which, in the context of Gabor analysis, leads to the so-called *Janssen representation* of the Gabor frame operator.

The proof of the theorem is based on two important features of the time-frequency plane $\mathbb{R}^d \times \widehat{\mathbb{R}}^d$.

(1) $\{U \mapsto \pi(\lambda)\, U\, \pi(\lambda)^* | \lambda \in \Lambda\}$ defines a unitary representation of Λ which gives the Λ-invariance of K.
(2) An application of the Poisson summation formula for the symplectic Fourier transform to $\sigma(P)$ with respect to the lattice Λ maps the periodization of

$$\sigma(K) = \sum_{\lambda \in \Lambda} T_\lambda(\sigma(P)) \tag{65}$$

to the sampling of the spreading function $\eta(P)$ on the lattice Λ°.

As an application we state that the Gabor frame operator $S_{g,\Lambda}$ of a Gabor system $\mathcal{G}(g, \Lambda)$ with $g \in S_0(\mathbb{R}^d)$ is generated by shifting a rank-one operator along the lattice Λ. In addition, we use the fact that the spreading function of a rank-one operator is the STFT. Altogether we therefore have

$$S_{g,\Lambda} = \frac{1}{|\Lambda|} \sum_{\lambda^\circ \in \Lambda^\circ} \langle g, \pi(\lambda^\circ)\gamma\rangle \pi(\lambda^\circ) \tag{66}$$

with $\gamma \in S_0(\mathbb{R}^d)$. The last equation (66) is the so-called Janssen representation of $S_{g,\Lambda}$ which decomposes $S_{g,\Lambda}$ into an *absolutely convergent* series of time-frequency shifts. In (66) we used implicitly another pleasant property of $S_0(\mathbb{R}^d)$.

Lemma 38: *Let $g, \gamma \in S_0(\mathbb{R}^d)$ and Λ a lattice in $\mathbb{R}^d \times \widehat{\mathbb{R}}^d$. Then (g, γ) satisfies Tolimieri-Orr's condition (A'):*

$$\sum_{\lambda \in \Lambda} |\langle g, \gamma_\lambda\rangle| < \infty, \qquad (A').$$

This stability of Condition (A') for $g, \gamma \in S_0(\mathbb{R}^d)$ with respect to a variation of the lattice makes Feichtinger's algebra such an important object in Gabor analysis. In a recent work Feichtinger and Kaiblinger have drawn some deep consequences from this fact. Roughly speaking, they proved that the set of functions in $S_0(\mathbb{R}^d)$ which generate a Gabor frame is *"open"* [15].

We close our discussion of the Gabor frame operator with a striking result of Gröchenig/Leinert on the quality of the canonical dual of a Gabor system $\mathcal{G}(g, \Lambda)$ generated by a window $g \in S_0(\mathbb{R}^d)$.

Theorem 39: *Let $g \in S_0(\mathbb{R}^d)$ and $\mathcal{G}(g, \Lambda)$ a Gabor frame of $L^2(\mathbb{R}^d)$. Then $\gamma_0 = S_{g,\Lambda}^{-1} g$ is in $S_0(\mathbb{R}^d)$.*

The proof is based on a noncommutative version of Wiener's lemma for the Banach algebra $\ell^1(\Lambda)$ with twisted convolution $\natural$ as product, and non-commutative involution $*$ as described above for the spreading function of a product of two operators in $\mathcal{L}(S_0, S_0')$ and the spreading function of the adjoint of an operator in $\mathcal{L}(S_0, S_0')$. A special case of their main result is that $(\ell^1, \natural, *)$ is a *symmetric* Banach algebra. In this context the Wiener lemma is expressed as the inverse-closedness of the Banach algebra

$$\mathcal{A}(\Lambda) = \{A \in \mathcal{B}(L^2(\mathbb{R}^d)) \,|\, A = \sum_{\lambda \in \Lambda} a_\lambda \pi(\lambda),\ (a_\lambda) \in \ell^1(\Lambda)\}$$

of absolutely convergent time-frequency series in the C^*-algebra $C^*(\Lambda)$ generated by time-frequency shifts $\{\pi(\lambda) : \lambda \in \Lambda\}$. In other words, the argument is based on the highly non-trivial fact that an element of $\mathcal{A}(\Lambda)$ which is invertible in $C^*(\Lambda)$ has its inverse already in $\mathcal{A}(\Lambda)$.

9. Conclusion and Outlook

It was the goal of this report to show how practical questions in signal processing lead to a very powerful mathematical model, the so-called Gabor frames, that induce highly structured operators which, in turn, open exciting viewpoints to advanced concepts in harmonic analysis. These concepts, such as the Segal algebra $S_0(\mathbb{R}^d)$, the Gelfand triple, and the spreading function provide deep insights into general properties of such a class of operators and relate Gabor analysis to other fields of physics and mathematics.

We have tried to give a consistent overview of the main tools of time-frequency analysis that are used to study signal expansions by means of Gabor systems. At the same time we also wanted to present state-of-the-art techniques for extracting the relevant parts of the models in order to draw a more transparent picture of the topic.

The central object of this paper is the Gabor frame-type operator. On one hand, it reduces to a sparse matrix in the finite dimensional case and can be treated by standard numerical methods. On the other hand, it is a special case of a class of Λ-periodic time-frequency operators that have a very attractive description in form of symbols over suitable function and distribution spaces. Specifically, the spreading function, the KN symbol and the Gelfand triple $(S_0, L^2, S_0')(\mathbb{R}^d)$ were introduced and studied to embed those results of the Gabor frame-type operator that initiated and popularized the Gabor theory in the mid nineties [9,29] in a larger framework of time-frequency analysis. The original ideas go back to the fundamental paper of Feichtinger and Kozek in 1998 [16]. These ideas have been evolving over time and other results, many of them nicely presented in [26], contributed to a better perception.

In the final part of the last section we returned to the central object, the Gabor frame-type operator, in order to stress the close connection between Gabor analysis and advanced time-frequency topics. The attentive reader, however, has certainly realized that it was not intended to wrap up all the relevant results. Indeed, the story is far from being complete and far from being ended. As a final taste in this sense, we shortly describe two more interesting topics. The first one is about so-called Gabor multipliers. The second one deals with the question of sampling time signals and studies Gabor expansions when increasing the sampling rate, a problem which has recently been analyzed by Kaiblinger.

In [18] Feichtinger and Nowak describe the foundation of a theory of (regular) *Gabor multipliers*, which are operators obtained by going from signal domain to some transform domain, and applying a pointwise multiplication operator before resynthesis. More generally speaking,

Definition 40: Assume we have $g_1, g_2 \in L^2(\mathbb{R}^d)$, Λ a lattice of $\mathbb{R}^d \times \widehat{\mathbb{R}}^d$ and let $\mathbf{m} = (m(\lambda))_{\lambda\in\Lambda}$ be a complex-valued sequence on Λ. Then the *Gabor multiplier* associated to the triple (g_1, g_2, Λ) with (upper) symbol $\mathbf{m}$ is defined by

$$G_{\mathbf{m}}(f) := G_{g_1,g_2,\Lambda,\mathbf{m}}(f) = \sum_{\lambda\in\Lambda} m(\lambda)\langle f, \pi(\lambda)g_1\rangle \pi(\lambda)g_2. \tag{67}$$

Therefore Gabor multipliers are infinite linear combinations of rank-one operators $f \mapsto \langle f, \pi(\lambda)g_1\rangle\pi(\lambda)g_2$ with coefficients given by $(m(\lambda))_{\lambda\in\Lambda}$. The function g_1 is called the *analysis window* and g_2 the *synthesis window* of the Gabor multiplier $G_{\mathbf{m}}$. A basic question arises naturally in this context. Namely, how the properties of the Gabor multiplier $G_{\mathbf{m}}$ depends on the decay of the multiplier sequence $(m(\lambda))_{\lambda\in\Lambda}$, the time-frequency concentration of g_1, g_2 and the time-frequency lattice Λ. In general g_1 and g_2 should be *Bessel atoms* with respect to the given lattice Λ and the strong symbol $\mathbf{m}$ is assumed to be bounded. In this case the coefficient mapping C_{g_1} using the analysis window g_1, mapping f to the sequence of sampling values $V_g f$ over Λ, is bounded from $L^2(\mathbb{R}^d)$ to $\ell^2(\Lambda)$, and also the synthesis mapping D_{g_2}, mapping a sequence $\mathbf{c} = (c(\lambda))_{\lambda\in\Lambda}$ to $\sum_{\lambda\in\Lambda} c(\lambda)\pi(\lambda)g_2$, is bounded from $\ell^2(\Lambda)$ to $L^2(\mathbb{R}^d)$ and thus the Gabor multiplier $G_{\mathbf{m}} = D_{g_2}C_{g_1}$ is bounded on $L^2(\mathbb{R}^d)$.

After our discussion of Gabor frame-type operators the interested reader will already conjecture that Gabor multipliers $G_{\mathbf{m}}$ with g_1 and g_2 have very nice properties. Furthermore, the terminology of Gelfand triples allows an elegant formulation of statements about Gabor multipliers. The following result is one of the main results in [18].

Theorem 41: *For every pair* (g_1, g_2) *in* $S_0(\mathbb{R}^d)$, *and any lattice* $\Lambda \in$

$\mathbb{R}^d \times \widehat{\mathbb{R}^d}$, the mapping from the multiplier $(m(\lambda))_{\lambda \in \Lambda}$ to the associated Gabor multiplier $G_{g_1,g_2,\mathbf{m},\Lambda}$ maps the Gelfand triple $(\ell^1, \ell^2, \ell^\infty)(\Lambda)$ into the bounded operators with kernel in the corresponding Gelfand triple $(S_0, L^2, S_0')(\mathbb{R}^d \times \widehat{\mathbb{R}}^d)$.

For more information about Gabor multipliers, e.g., their eigenvalue behavior, and their relation to time-varying filters we refer the reader to [18].

The work [31] of Kaiblinger on the approximation of the dual Gabor window γ of a Gabor frame $\mathcal{G}(g, \Lambda)$ in $L^2(\mathbb{R}^d)$ is interesting for two reasons: On one hand, it shows that our study of finite-dimensional Gabor frames is not only a motivation for the treatment of Gabor frames for $L^2(\mathbb{R}^d)$. On the other hand, all his results about approximation of a continuous function by finite methods only work for a Gabor atom g in $S_0(\mathbb{R}^d)$.

Kaiblinger's result are based on a synthesis of fast algorithms for the computation of dual window of a Gabor frame in $\mathbb{C}^n$ and on the fact that these Gabor frames for $\mathbb{C}^n$ are obtained in a simple way from the original Gabor frame $\mathcal{G}(g, \Lambda)$. He also showed that the approximate dual windows converge not just in $L^2(\mathbb{R}^d)$ but indeed in $S_0(\mathbb{R}^d)$, which implies the convergence of the corresponding frame operators in the operator norm on $L^2(\mathbb{R}^d)$. For further information we refer to [31].

References

1. R. Balan, P. Casazza, C. Heil, and Z. Landau. Deficits and excesses of frames. *Adv. Comput. Math.*, 18(2-4):93–116, 2003.
2. R. Balan, P. Casazza, C. Heil, and Z. Landau. Excesses of Gabor frames. *Appl. Comput. Harmon. Anal.*, 14(2):87–106, 2003.
3. R. Balian. Un principe d'incertitude fort en théorie du signal on en mécanique quantique. *C. R. Acad. Sci. Paris*, 292:1357–1362, 1981.
4. D. Bau and L. N. Trefethen. *Numerical Linear Algebra.* Philadelphia, PA: SIAM, 2000.
5. O. Christensen. *An Introduction to Frames and Riesz Bases.* Birkhäuser, 2003.
6. O. Christensen and Y. C. Eldar. Oblique dual frames and shift-invariant spaces. *Appl. and Comp. Harm. Anal.*, 17(1):48–68, 2004.
7. S. Dahlke, M. Fornasier, and T. Raasch. Adaptive frame methods for elliptic operator equations. to appear in Adv. Comp. Math., 2005.
8. I. Daubechies, S. Jaffard, and J. L. Journé. A simple Wilson orthonormal basis with exponential decay. *SIAM J. Math. Anal.*, 22(2):554–572, 1991.

9. I. Daubechies, H. J. Landau, and Z. Landau. Gabor time-frequency lattices and the Wexler-Raz Identity. *J. Four. Anal. and Appl.*, 1(4):437–478, 1995.
10. M. Dörfler, H. G. Feichtinger, and K. Gröchenig. Time-frequency partitions for the Gelfand triple (S_0, L^2, S_0'). Submitted, 2004.
11. R. Duffin and A. Schaeffer. A class of nonharmonic Fourier series. *Trans. Amer. Math. Soc.*, 72:341–366, 1952.
12. H. G. Feichtinger, *On a New Segal Algebra, Monatsh. Math.*, 92:269-289, 1981.
13. H. G. Feichtinger and K. Gröchenig. Banach spaces related to integrable group representations and their atomic decompositions. I. *J. Funct. Anal.* 86, No.2: 307–340, 1989.
14. H. G. Feichtinger, K. Gröchenig, and D. Walnut. Wilson bases and modulation spaces. *Math. Nachrichten*, 155:7–17, 1992.
15. H. G. Feichtinger and N. Kaiblinger. Varying the time-frequency lattice of Gabor frames. *Trans. Am. Math. Soc.*, 356(5):2001–2023, 2004.
16. H. G. Feichtinger and W. Kozek. Quantization of TF–lattice invariant operators on elementary LCA groups. In H. G. Feichtinger and T. Strohmer, editors, *Gabor Analysis and Algorithms: Theory and Applications*, chapter 7, pages 233–266. Birkhäuser, Boston, 1998.
17. H. G. Feichtinger and F. Luef. Wiener amalgam spaces for the fundamental identity of Gabor analysis. *Collect. Math.*, 57 (Extra Volume): 233–253, 2006.
18. H. G. Feichtinger and K. Nowak. A first survey of Gabor multipliers. In H. G. Feichtinger and T. Strohmer, editors, *Advances in Gabor analysis*, Applied and Numerical Harmonic Analysis, pages 99–128. Birkhäuser, Boston, 2003.
19. H. G. Feichtinger and G. Zimmermann. A Banach space of test functions for Gabor analysis. In H. G. Feichtinger and T. Strohmer, editors, *Gabor Analysis and Algorithms: Theory and Applications*, pages 123–170. Birkhäuser, Boston, 1998.
20. G. B. Folland. *A course in Abstract Harmonic Analysis.* Studies in Advanced Mathematics. Boca Raton, FL: CRC Press, 1995.
21. D. Gabor. Theory of communication. *Proc. IEE (London)*, 93(III):429–457, November, 1946.
22. I. M. Gelfand and A. G. Kostyuchenko. Entwicklung nach Eigenfunktionen von Differentialoperatoren und anderen Operatoren. *Dokl. Akad. Nauk SSSR*, 103:349–352, 1955.
23. I. M. Gelfand and N. Ya. Vilenkin. *Applications of Harmonic Analysis.* New York and London: Academic Press. XIV, 1964.
24. G. H. Golub and C. F. van Loan. *Matrix Computations.* John Hopkins University Press, Baltimore, 1989.
25. K. Gröchenig. An uncertainty principle related to the Poisson summation formula. *Stud. Math.*, 121(1):87–104, 1996.
26. K. Gröchenig. *Foundations of Time-Frequency Analysis.* Birkhäuser, 2001.
27. K. Gröchenig and M. Leinert. Wiener's Lemma for twisted convolution and Gabor frames. *J. Amer. Math. Soc.*, 17(1):1–18, 2003.
28. C. Heil. A basis theory primer. Unpublished Manuscript, July 1997.

29. A. J. E. M. Janssen. Duality and biorthogonality for Weyl-Heisenberg frames. *J. Four. Anal. and Appl.*, 1(4):403–436, 1995.
30. A. J. E. M. Janssen. Classroom proof of the density theorem for Gabor systems. Unpublished Manuscript, 2005.
31. N. Kaiblinger. Approximation of the Fourier transform and the dual Gabor window. *J. Fourier Anal. Appl.*, 11(1):25–42, 2005.
32. J. G. Kirkwood. Quantum statistics of almost classical assemblies. *Phys. Rev.*, II. Ser. 44:31–37, 1933.
33. S. Li and H. Ogawa. Pseudoframes for subspaces with applications. *J. Four. Anal. and Appl.*, 10(4):409–431, 2004.
34. E. H. Lieb. Integral bounds for radar ambiguity functions and Wigner distributions. *J. Math. Phys.*, 31(3):594–599, 1990.
35. F. Low. Complete sets of wave packets. In C. DeTar, editor, *A Passion for Physics - Essay in Honor of Geoffrey Chew*, pages 17–22. World Scientific, Singapore, 1985.
36. Y. I. Lyubarskii. Frames in the Bargmann space of entire functions. *Adv.Soviet Math.*, 429:107–113, 1992.
37. S. Qiu and H. G. Feichtinger. Discrete Gabor structures and optimal representation. *IEEE Trans. Signal Proc.*, 43(10):2258–2268, 1995.
38. A. Ron and Z. Shen. Frames and stable bases for subspaces in $L^2(\mathbb{R}^d)$: The duality principle of Weyl-Heisenberg sets. In M. Chu, D. Brown and D. Ellison, editors, *Proceedings of the Lanczos Centenary Conference Raleigh, NC*, pages 422–425. SIAM Pub., 1993.
39. A. Ron and Z. Shen. Weyl-Heisenberg frames and Riesz basis on $L^2(\mathbb{R}^d)$. *Duke Math. J.*, 89(2):273–282, 1997.
40. W. Rudin. *Functional Analysis.* Mc Graw-Hill, 2nd edition, 1991.
41. K. Seip and R. Wallsten. Density theorems for sampling and interpolation in the Bargmann-Fock space II. *J. Reine Angewandte Mathematik*, 429:107–113, 1992.
42. T. Strohmer. Numerical algorithms for discrete Gabor expansions. In H. G. Feichtinger and T. Strohmer, editors, *Gabor Analysis and Algorithms: Theory and Applications*, pages 267–294. Birkhäuser, Boston, 1998.
43. J. Wexler and S. Raz. Discrete Gabor expansions. *Signal Processing*, 21(3):207–220, 1990.
44. E. P. Wigner. On the quantum correction for thermo-dynamic equilibrium. *Phys. Rerv. Letters*, 40:749–759, 1932.
45. W. Y. Zou and Y. Wu. COFDM: An overview. *IEEE Trans. Broadc.*, 41(1):1–8, 1995.

SOME ITERATIVE ALGORITHMS TO COMPUTE CANONICAL WINDOWS FOR GABOR FRAMES

A. J. E. M. Janssen

Philips Research Laboratories WO-02
5656 AA Eindhoven, The Netherlands
E-mail: a.j.e.m.janssen@philips.com

We analyze some iterative algorithms for the computation of the canonical tight window g^t and the canonical dual window g^d associated with a Gabor frame (g, a, b). As to the computation of g^t, we consider algorithms that do require inversion of intermediate frame operators as well as algorithms that do not require inversions. As to the computation of g^d, we naturally consider algorithms where no frame operator inversions are required. These algorithms have safe but conservative versions, with guaranteed convergence of prescribed order but with suboptimal convergence constants, and smart but risky versions, with near-optimal convergence of prescribed order which is, however, guaranteed only if the frame bound ratio A/B of (g, a, b) exceeds an analytically given lower bound. Thus we propose for g^t an algorithm, using inversions, with quadratic convergence, and two algorithms, using no inversions, with quadratic and cubic convergence, respectively, and we identify for these algorithms the safe and the smart versions. For g^d we propose two algorithms, without inversions, with quadratic and cubic convergence, respectively, and also for these algorithms we identify the safe and the smart versions. All these algorithms can be formulated by using a general mechanism for proposing the recursion step in an approximation scheme for g^t and g^d with a prescribed error decay. The tools used to analyze the algorithms are the calculus of frame operators, the spectral mapping theorem, and Kantorovich's inequality.

1. Introduction

We consider for $a > 0$, $b > 0$ and $g \in L^2(\mathbb{R})$ the Gabor system

$$(g, a, b) = (g_{na,mb})_{n,m \in \mathbb{Z}} \;, \tag{1}$$

where for $x, y \in \mathbb{R}$ we denote by

$$g_{x,y}(t) = e^{2\pi iyt}\, g(t-x)\ , \qquad t \in \mathbb{R}\ , \tag{2}$$

the time-frequency shifted version $g_{x,y}$ of g. We assume that (g, a, b) is a frame (so that it is assumed implicitly that $ab \le 1$). We denote by S the frame operator

$$S : f \in L^2(\mathbb{R}) \to \sum_{n,m=-\infty}^{\infty} (f, g_{na,mb})\, g_{na,mb} \in L^2(\mathbb{R})\ , \tag{3}$$

and we denote by A, B the best lower, upper frame bounds

$$A = \min \sigma(S) > 0\ , \qquad B = \max \sigma(S) < \infty\ , \tag{4}$$

where $\sigma(S)$ is the spectrum of the positive, bounded operator S. We, furthermore, denote by g^t and g^d the canonical tight window and the canonical dual window associated with the frame (g, a, b) according to

$$g^t = S^{-1/2} g\ , \qquad g^d = S^{-1} g\ , \tag{5}$$

respectively.

We refer for extensive information on Gabor frames and the role of the canonical windows to [6], [1], Ch. 4, [2], [4], Chs. 5–9, and to Feichtinger's contribution [3] to the present IMS series volume. In particular, we have that S, $S^{-1/2}$, S^{-1}, etc. commute with the relevant shift operators $f \in L^2(\mathbb{R}) \to f_{na,mb} \in L^2(\mathbb{R})$ with integer n, m. Furthermore, the triples (g^t, a, b) and (g^d, a, b) are themselves Gabor frames, with respective frame operators I and S^{-1}.

2. Overview

In this section we present a number of iterative algorithms to approximate g^t and g^d, and we describe the results obtained for these algorithms in this contribution. We let $\gamma_0 = g$ and for $k = 0, 1, \ldots$, we set

$$\text{I.} \quad \gamma_{k+1} = \tfrac{1}{2}\,\alpha_k \gamma_k + \tfrac{1}{2}\,\beta_k S_k^{-1} \gamma_k\ , \tag{6}$$

$$\text{II.} \quad \gamma_{k+1} = \tfrac{3}{2}\,\varepsilon_{k0} \gamma_k - \tfrac{1}{2}\,\varepsilon_{k1} S_k \gamma_k\ , \tag{7}$$

$$\text{III.} \quad \gamma_{k+1} = \tfrac{15}{8}\,\varepsilon_{k0} \gamma_k - \tfrac{5}{4}\,\varepsilon_{k1} S_k \gamma_k + \tfrac{3}{8}\,\varepsilon_{k2} S_k^2 \gamma_k\ , \tag{8}$$

$$\text{IV.} \quad \gamma_{k+1} = 2\delta_{k0} \gamma_k - \delta_{k1} S_k g\ , \tag{9}$$

$$\text{V.} \quad \gamma_{k+1} = 3\delta_{k0} \gamma_k - 3\delta_{k1} S_k g + \delta_{k2} S S_k \gamma_k\ , \tag{10}$$

respectively. Here S_k is the frame operator corresponding to (γ_k, a, b), and the scalars $\alpha_k, \beta_k > 0$ in (6), ε_{ki} in (7)–(8) and δ_{ki} in (9)–(10) have to be chosen appropriately. In particular, it will be necessary to choose these scalars in such a way that (γ_{k+1}, a, b) is again a Gabor frame.

The algorithm with recursion step I and with

$$\alpha_k = \|\gamma_k\|^{-1}, \qquad \beta_k = \|S_k^{-1}\gamma_k\| \tag{11}$$

has been analyzed in [6]. It is shown there that there is quadratic and monotone convergence (in a sense that will be clear to the reader after Sec. 3) of $\gamma_k/\|\gamma_k\|$ to $g^t/\|g^t\|$. Before we proceed it should be said that [6] contains a number of rather innocent but disturbing errors; a corrected version of [6] can be found at

`http://www.math.ucdavis.edu/~strohmer/papers/2000/tight.html`

In Sec. 4 we shall repeat the arguments given in [6].

The algorithms with recursion step II and IV will be analyzed in detail in the present contribution; they yield sequences γ_k with $\gamma_k/\|\gamma_k\| \to g^t/\|g^t\|$ and $\gamma_k/\|\gamma_k\| \to g^d/\|g^d\|$, respectively, when ε_{k0}, ε_{k1} and δ_{k0}, δ_{k1} are chosen appropriately, and the convergence is (at least) quadratic. These algorithms have safe but conservative versions and smart but risky versions, depending on the way ε_{k0}, ε_{k1} and δ_{k0}, δ_{k1} are chosen.

The safe but conservative versions are obtained by replacing

$$g \text{ by } g/\hat{B}^{1/2}, \qquad S \text{ by } S/\hat{B}, \tag{12}$$

where $\hat{B}$ is any number $\geq B = \max\sigma(S)$, and by choosing $\varepsilon_{k0} = \varepsilon_{k1} = 1$ and $\delta_{k0} = \delta_{k1} = 1$ in II and IV, respectively. We shall show that in either case II or IV there is quadratic and monotone convergence in these safe modes, no matter how small the frame bound ratio A/B of the initial frame (g, a, b) is as long as it is positive.

The smart but risky modes are obtained by leaving g and S as they were and by taking ε_{k0}, ε_{k1} and δ_{k0}, δ_{k1} such that $\varepsilon_{k0}/\varepsilon_{k1}$ and δ_{k0}/δ_{k1} are between A_k and B_k, where A_k and B_k are the best lower and upper frame bound of the current frame operator S_k. We shall consider the choice

$$\varepsilon_{k0} = \|\gamma_k\|^{-1}, \qquad \varepsilon_{k1} = \|S_k\gamma_k\|^{-1} \tag{13}$$

and

$$\delta_{k0} = \|\gamma_k\|^{-1}, \qquad \delta_{k1} = \|S_k g\|^{-1}, \tag{14}$$

that yield $\varepsilon_{k0}/\varepsilon_{k1} \in [A_k, B_k]$ and $\delta_{k0}/\delta_{k1} \in [A_k, B_k]$, indeed. We shall show that for the smart version of II so obtained we have quadratic and monotone convergence of $\gamma_k/\|\gamma_k\|$ to $g^t/\|g^t\|$ when the initial frame bound ratio $A/B > \frac{1}{2}$, and that for the smart version of IV so obtained we have quadratic and monotone convergence of $\gamma_k/\|\gamma_k\|$ to $g^d/\|g^d\|$ when the initial frame bound ratio $A/B > \frac{1}{2}(\sqrt{5}-1)$.

The constants involved in the smart versions of II and IV that describe the convergence speed are considerably better than those involved in the safe modes. An important point is that one can switch from the safe to the smart mode (and vice versa) without changing the limiting window. This gives the opportunity to start in the safe mode, and to change to the smart mode as soon as one is confident that the frame bound ratio A_k/B_k is large enough.

The algorithms with recursion steps III and V are analyzed in detail in [7], especially with respect to convergence behavior in the smart modes. Also in these cases there are safe but conservative modes, and one can switch freely from safe to smart modes, and vice versa.

It is not the purpose of this contribution to compare the performance of the presented algorithms with existing ones in the literature; we just give an indication of the performance. Preliminary experiments, with the standard Gaussian window $g(t) = 2^{1/4}\exp(-\pi t^2)$ and $a = b = 1/\sqrt{2}$, have shown that the smart versions produce 10^{-15} accurate approximations within 4–7 steps for the algorithms with quadratic convergence and within 2 or 3 steps for the algorithms with cubic convergence. The number of steps required for the safe versions is typically a factor $1\frac{1}{2}$–2 larger.

This contribution is organized as follows. In Sec. 3 we shall present our main tools to analyze the algorithms I–V. These are the spectral mapping theorem to relate the best frame bounds A_{k+1}, B_{k+1} of (γ_{k+1}, a, b) to the best frame bounds A_k, B_k of (γ_k, a, b), the Kantorovich inequality for transferring convergence of frame operators to convergence of the corresponding windows and, finally, some elementary inequalities to produce numbers (like $\varepsilon_{k0}/\varepsilon_{k1}$ and δ_{k0}/δ_{k1}) that are guaranteed to lie between A_k and B_k. In Sec. 4 the effectiveness of these tools for establishing convergence results is demonstrated by redoing the analysis for the algorithm with recursion step I. Next, in Sec. 5, we present a rationale for proposing iterative algorithms of the types II-III and IV–V meant for approximation of g^t and g^d, respectively, with arbitrary order of convergence m. Such a rationale is badly needed since until now nothing has been said about what has led us to write down the recursions II–V with the constants and terms at the right-hand sides that are characteristic for any of these recursions. In Sec. 6 we shall give a detailed analysis of algorithm II using the tools developed in Sec. 3 so as to derive the results for II in safe and smart mode as announced above. In Sec. 7 we shall do a similar effort as in Sec. 6, but now for the algorithm with recursion step IV. Finally, in Sec. 8 we shall announce a number of results from [7]; these include comments on the algorithms III and IV with (conditional) cubic convergence together with an appropriately sharpened version of Kantorovich's inequality.

We conclude this overview by some comments of more historical and/or motivational nature. The algorithm I was proposed by Feichtinger and Strohmer (independently from one another) around 1995 and analyzed, see [6], around 2000. The intuitive idea here was that $S^{-1/2}g = g^t$ is about halfway between g and $S^{-1}g = g^d$ and that this should be reflected in some sense by a recursion approximating it. Here the occurrence of inverses of frame operators should be considered as much less a problem from a numerical point of view than having to form inverse square roots of frame operators. The algorithm IV was proposed to the author by Feichtinger in Vienna, December 2001; apparently, not much can be said about the motivation of it, except that some adequate guess and try work as well as a good feeling for iterative methods has been instrumental here. The algorithms II and the algorithms III, V were proposed by the author himself just before and just after December 2001, respectively. All these inputs lead to the work [7] in which the algorithms II–V were analyzed in their smart modes. The modification of algorithms from smart (but risky) modes to safe modes was prompted by an observation of Hampejs in the spring of 2002, [5], where a scaling of the window g and the frame operator S somewhat similar to (12) and constant ε's and δ's in II and IV, respectively, gave rise to versions of algorithms II and IV converging under less restrictive conditions on A/B than those required in the smart modes considered in [7]. Finally, the author was informed of recent work of M. Lammers [9] in which Newton type methods for the approximation of tight and dual windows for a Gabor frame (g, a, b) were proposed. The approach in [9] depends on a convolution product $\odot_\alpha$ for windows, in which a pre-chosen tight window α occurs, and yields, in general, non-canonical tight and dual windows (without restrictions on the initial frame bounds A, B).

3. Basic Tools

In this section we present the basic tools by which we analyze the algorithms presented in Sec. 2. We concentrate here on algorithms of type I–III to approximate g^t, and we present details for how to use the tools to analyze algorithms of the type IV–V to approximate g^d in Subsec. 5.2.

In the case of algorithms I-III we have recursion steps of the type $\gamma_{k+1} = \varphi_k(S_k)\,\gamma_k$ with φ_k an analytic function. Since γ_k is supposed to approximate g^t and since the frame operator corresponding to (g^t, a, b) is the identity I, the frame bound ratio A_k/B_k of the frame (γ_k, a, b) should approximate 1. The following theorem can be used to relate the frame bounds A_{k+1}, B_{k+1} of $(\gamma_{k+1} = \varphi_k(S_k)\,\gamma_k, a, b)$ to the frame bounds A_k, B_k of (γ_k, a, b); it thus serves as a tool to track the quantity A_k/B_k during the iteration.

Theorem 1: *Let (g, a, b) be a Gabor frame with frame operator S and let φ be analytic in an open neighborhood of $\sigma(S)$. Moreover, assume that φ is positive on $\sigma(S)$. Then $(\varphi(S)\,g, a, b)$ is a Gabor frame with frame operator $S\,\varphi^2(S)$ and best lower and upper frame bounds given by*

$$\min_{s\in\sigma(S)} s\,\varphi^2(s) \quad \text{and} \quad \max_{s\in\sigma(S)} s\,\varphi^2(s)\ , \tag{15}$$

respectively. Furthermore,

$$(\varphi(S)\,g)^t = g^t\ , \tag{16}$$

i.e. g and $\varphi(S)\,g$ have the same canonical tight window for the shift parameters a, b.

Proof: For $f \in L^2(\mathbb{R})$ we compute, see the definition (3) of frame operator,

$$\begin{aligned}
&\sum_{n,m} (f, (\varphi(S)\,g)_{na,mb})(\varphi(S)\,g)_{na,mb} \\
&= \sum_{n,m} (f, \varphi(S)\,g_{na,mb})\,\varphi(S)\,g_{na,mb} \\
&= \sum_{n,m} (\varphi(S)\,f, g_{na,mb})\,\varphi(S)\,g_{na,mb} \\
&= \varphi(S) \sum_{n,m} (\varphi(S)\,f, g_{na,mb})\,g_{na,mb} = \varphi(S)\,S\,\varphi(S)\,f\ ,
\end{aligned} \tag{17}$$

showing that the frame operator of $(\varphi(S)\,g, a, b)$ equals $S\,\varphi^2(S)$. In (17) it has been used that S, and hence $\varphi(S)$, commutes with all relevant frame operators and that $\varphi(S)$ is a bounded positive operator of $L^2(\mathbb{R})$. By the spectral mapping theorem, we have that $\sigma(\varphi(S)) = \varphi(\sigma(S))$, and this implies that the best frame bounds of $(\varphi(S)\,g, a, b)$ are given by the numbers in (15).

As to (16) we observe that $S\,\varphi^2(S)$ is positive and that, see (5),

$$(\varphi(S)\,g)^t = (S\,\varphi^2(S))^{-1/2}\,\varphi(S)\,g = S^{-1/2}g = g^t\ . \tag{18}$$

This completes the proof. □

Note: The first half of Theorem 1 is also true when φ is just real, rather than positive, on $\sigma(S)$, except that $(\varphi(S)g, a, b)$ does not need to be a frame.

For the iterations of the type $\gamma_{k+1} = \varphi_k(S_k)\,\gamma_k$ that we consider in algorithms I-III there is the following consequence of Theorem 1 and the fact that $\sigma(S_k) \subset [A_k, B_k]$.

Corollary 2: *Assume that g is as in Theorem 1 and that $(\gamma_k)_{k=0,1,\ldots}$ is a sequence in $L^2(\mathbb{R})$ generated as $\gamma_0 = g$, $\gamma_{k+1} = \varphi_k(S_k)\,\gamma_k$, $k = 0, 1, \ldots$, where S_k is the frame operator corresponding to (γ_k, a, b) and φ_k is analytic around and positive on $\sigma(S_k)$. Then*

$$A_{k+1} = \min_{s\in\sigma(S_k)} s\,\varphi_k^2(s) \geq \min_{s\in[A_k,B_k]} s\,\varphi_k^2(s)\ , \tag{19}$$

$$B_{k+1} = \max_{s\in\sigma(S_k)} s\,\varphi_k^2(s) \leq \max_{s\in[A_k,B_k]} s\,\varphi_k^2(s)\ , \tag{20}$$

and

$$\gamma_k^t = S_k^{-1/2}\gamma_k = S^{-1/2}g = g^t \tag{21}$$

for all $k = 0, 1, \ldots$.

With Corollary 2 in hand, we can monitor the frame bound ratio $Q_k = A_k/B_k$. It can be expected that γ_k is close to g^t when A_k/B_k is close to 1. This can be made more explicit by using the following inequalities. Let T be a positive linear operator of a Hilbert space H and let $C = \min\sigma(T) > 0$, $D = \max\sigma(T) < \infty$. Then there holds

$$\frac{2CD}{C^2+D^2} \leq \frac{\|Tf\|^2}{\|f\|\,\|T^2f\|} \leq 1\ , \qquad f \in H\ ; \tag{22}$$

the first inequality here is due to Kantorovich, see [8].

Theorem 3: *Assume that γ_k is as in Corollary 2. Then*

$$\left\|\frac{\gamma_k}{\|\gamma_k\|} - \frac{g^t}{\|g^t\|}\right\| \leq (1 - Q_k^{1/4})\sqrt{\frac{2}{1+Q_k^{1/2}}}\ , \tag{23}$$

where $Q_k = A_k/B_k$.

Proof: We have by (21) that $g^t = S_k^{-1/2}\gamma_k$. Hence

$$\Big\|\frac{\gamma_k}{\|\gamma_k\|} - \frac{g^t}{\|g^t\|}\Big\|^2 = 2\Big(1 - \mathrm{Re}\Big[\frac{(\gamma_k, g^t)}{\|\gamma_k\|\,\|g^t\|}\Big]\Big)$$

$$= 2\Big(1 - \frac{\|S_k^{-1/4}\gamma_k\|^2}{\|\gamma_k\|\,\|S_k^{-1/2}\gamma_k\|}\Big)$$

$$\leq 2\Big(1 - \frac{2B_k^{-1/4}A_k^{-1/4}}{B_k^{-1/2} + A_k^{-1/2}}\Big)$$

$$= 2\,\frac{(B_k^{1/4} - A_k^{1/4})^2}{A_k^{1/2} + B_k^{1/2}} = 2\,\frac{(1 - Q_k^{1/4})^2}{1 + Q_k^{1/2}}\,. \tag{24}$$

Here we have used the first inequality in (22) with $T = S_k^{-1/2}$ and $C = B_k^{-1/4}$, $D = A_k^{-1/4}$ while choosing $f = \gamma_k$. This completes the proof. □

Note: The monotonicity statements, as occur in the formulation of the results in Sec. 2, refer to the monotonicity of the Q_k and not to the monotonicity of the quantities at the left-hand side of (23).

In the smart versions of the algorithms the availability of numbers between the best frame bounds A_k and B_k, without knowing A_k and B_k themselves, is required. For this we have the following result.

Theorem 4: *Let (g, a, b) be a Gabor frame with frame operator S and best frame bounds A, B. Then for any $h \in L^2(\mathbb{R})$ there holds*

$$A \leq \frac{\|h\|}{\|S^{-1}h\|} \leq \frac{\|Sh\|}{\|h\|} \leq B\,. \tag{25}$$

Furthermore, we have

$$A \leq \frac{\|g\|}{\|S^{-1}g\|} \leq \frac{\|g\|^2}{(S^{-1}g, g)} = \frac{1}{ab}\,\|g\|^2 \leq \frac{(Sg, g)}{\|g\|^2} \leq \frac{\|Sg\|}{\|g\|} \leq B\,. \tag{26}$$

Proof: These inequalities are elementary, and the identity

$$\frac{\|g\|^2}{(S^{-1}g, g)} = \frac{1}{ab}\,\|g\|^2 \tag{27}$$

in the middle of the chain in (26) follows from $(g, g^d) = (g, S^{-1}g) = ab$ (Wexler-Raz). This completes the proof. □

4. Analysis of Recursion I to Approximate g^t

In this section we redo the analysis given in [6], Sec. 4 to demonstrate the effectiveness of the tools developed in Sec. 3 to analyze the algorithm with I as recursion step. Hence we consider the recursion

$$\gamma_0 = g \; ; \qquad \gamma_{k+1} = \tfrac{1}{2}\,\alpha_k \gamma_k + \tfrac{1}{2}\,\beta_k S_k^{-1}\gamma_k\,, \quad k = 0, 1, \dots\,, \tag{28}$$

where $\alpha_k > 0$, $\beta_k > 0$, and we pay special attention to the case that

$$\alpha_k = \|\gamma_k\|^{-1}\,, \qquad \beta_k = \|S_k^{-1}\gamma_k\|^{-1}\,. \tag{29}$$

We have in the present case that

$$\gamma_{k+1} = \varphi_k(S_k)\,\gamma_k \; ; \qquad \varphi_k(s) = \tfrac{1}{2}\,\alpha_k + \tfrac{1}{2}\,\beta_k s^{-1}\,, \tag{30}$$

where we observe that φ_k is analytic and positive on $(0, \infty)$. Hence by Corollary 2 there holds

$$A_{k+1} \geq \min_{s\in[A_k, B_k]} s\,\varphi_k^2(s)\,, \qquad B_{k+1} \leq \max_{s\in[A_k, B_k]} s\,\varphi_k^2(s)\,, \tag{31}$$

and thus

$$Q_{k+1} = \frac{A_{k+1}}{B_{k+1}} \geq \frac{\min\, s\,\varphi_k^2(s)}{\max\, s\,\varphi_k^2(s)}\,, \tag{32}$$

where min and max are taken over $[A_k, B_k]$.

In Fig. 1 we have plotted for a fixed value of $\alpha > 0$, $\beta > 0$ the graph of the mapping

$$s > 0 \to s\,\varphi^2(s) \; ; \qquad \varphi(s) = \tfrac{1}{2}\,\alpha + \tfrac{1}{2}\,\beta\, s^{-1}\,, \tag{33}$$

and we consider the minimum and maximum of $s\,\varphi^2(s)$ over an interval $[A, B]$ where we distinguish between the cases

$$\text{(i) } A \leq \frac{\beta}{\alpha} \leq B\,, \qquad \text{(ii) } A \leq B < \frac{\beta}{\alpha}\,, \qquad \text{(iii) } \frac{\beta}{\alpha} < A \leq B\,. \tag{34}$$

It is thus seen that

$$\text{Case (i)}: \quad \frac{\text{min over } [A, B]}{\text{max over } [A, B]} = \frac{\alpha\beta}{\text{max over } \{A, B\}}\,, \tag{35}$$

$$\text{Case (ii)}: \quad \frac{\text{min over } [A, B]}{\text{max over } [A, B]} = \frac{B\,\varphi^2(B)}{A\,\varphi^2(A)}\,, \tag{36}$$

$$\text{Case(iii)}: \quad \frac{\text{min over } [A, B]}{\text{max over } [A, B]} = \frac{A\,\varphi^2(A)}{B\,\varphi^2(B)}\,. \tag{37}$$

An elementary calculation shows that the three right-hand sides of (35)–(37) are all $> A/B$.

Returning to (32) we see that $Q_{k+1} > Q_k$, no matter what A_k and B_k are as long as $A_k > 0$, $B_k < \infty$. Hence, in each step of the recursion in (28) the frame (γ_k, a, b) gets tighter. Furthermore, this process of tightening up is seen to be fastest when we consistently choose α_k, β_k such that case (i) in (34) occurs with α_k, β_k, A_k, B_k instead of α, β, A, B. In the latter case it can be shown by elementary means that

$$\frac{4Q_k}{(1+Q_k)^2} \le \frac{\alpha_k \beta_k}{\max\{A_k \varphi_k^2(A_k), B_k \varphi_k^2(B_k)\}} \le \frac{4Q_k^{1/2}}{(1+Q_k^{1/2})^2} \,. \tag{38}$$

Equality in the left-hand side inequality in (38) occurs if and only if β_k/α_k equals A_k or B_k, and equality in the right-hand side inequality in (38) occurs if and only if $\beta_k/\alpha_k = (A_k B_k)^{1/2}$. Thus we certainly have that

$$Q_{k+1} \ge \frac{4Q_k}{(1+Q_k)^2} = 1 - \left(\frac{1-Q_k}{1+Q_k}\right)^2 , \tag{39}$$

and this implies at least quadratic convergence of Q_k to 1. Then we can use Theorem 3 in Sec. 3 to transfer convergence of frame operators to windows.

The convergence of Q_k to 1 can be considerably faster than what (39) would give with the equality sign. This is, for instance, the case when $\sigma(S)$ is a small subset of $[A, B]$, so that $\sigma(S_k)$ is a small subset of $[A_k, B_k]$. We note that Q_{k+1} equals the middle quantity in (38) when $\sigma(S) = [A, B]$ (so that $\sigma(S_k) = [A_k, B_k]$). Another instance where a considerable sharpening of (39) occurs is the case that we manage to choose α_k, β_k such that β_k/α_k is close to $(A_k B_k)^{1/2}$ so that near-equality in the second inequality in (38) occurs. Noting that for Q close to 1

$$\left(\frac{1-Q}{1+Q}\right)^2 \approx \tfrac{1}{4}\,(1-Q)^2 \,, \qquad \left(\frac{1-Q^{1/2}}{1+Q^{1/2}}\right)^2 \approx \tfrac{1}{16}\,(1-Q)^2 \,, \tag{40}$$

we see that the convergence constant $\frac{1}{4}$ is improved to $\frac{1}{16}$ when we succeed in choosing β_k/α_k close to $(A_k B_k)^{1/2}$.

The condition that $\beta_k/\alpha_k \in [A_k, B_k]$ can be satisfied in many ways as is evident from Theorem 4. In [6], Sec. 4 the choice (29) has been made, which corresponds to the second term in the chain in (26), so that both members at the right-hand side of the recursion step in (28) have norm $\frac{1}{2}$.

We now point at the fact that there is quadratic convergence in (28), even when one does not pay special attention in choosing α_k, β_k. This is the case, for instance, when one chooses $\alpha_k = \beta_k = 1$ for all k. Consider this case in the event that case (iii) in (37) holds. Then there holds, see Fig. 1 (iii),

$$A_{k+1} = 1 + \frac{(A_k - 1)^2}{4A_k} \ , \qquad B_{k+1} = 1 + \frac{(B_k - 1)^2}{4B_k} \ , \tag{41}$$

and from this one sees that both A_k and B_k tend to 1 quadratically (equality in (41) since the best frame bounds A_k, B_k are in $\sigma(S_k)$). A similar thing occurs when case (i) or (ii) holds in (34). Evidently, this careless strategy comes at a price when A, B are both very small or very large. And also the asymptotic convergence constant is normally not better than $\frac{1}{4}$, where a potential $\frac{1}{16}$ could be obtained when one is successful in choosing β_k/α_k somewhere in the middle of the interval $[A_k, B_k]$.

5. Proposing Iterations Without Inversions

In this section we develop a rationale for proposing iterative algorithms for the approximation of g^t or g^d using no frame operator inversions and with a given order m of convergence. This yields the algorithms II–III with

$$\varepsilon_{k0} = \|\gamma_k\|^{-1} \ , \qquad \varepsilon_{k1} = \|S_k\gamma_k\|^{-1} \ , \qquad \varepsilon_{k2} = \|S_k^2\gamma_k\| \ , \tag{42}$$

and algorithms IV–V with

$$\delta_{k0} = \|\gamma_k\|^{-1} \ , \qquad \delta_{k1} = \|S_k g\|^{-1} \ , \qquad \delta_{k2} = \|SS_k\gamma_k\|^{-1} \ , \tag{43}$$

respectively, corresponding to the cases $m = 2$ and 3. In Subsec. 5.1 we consider this rationale for iterations meant to approximate g^t. In Subsec. 5.2 we consider this issue for the approximation of g^d, and there we first show how the tools developed in Sec. 3 are to be used for analyzing iterations for g^d.

5.1. *Iterations for* g^t

Let $m = 2, 3, \ldots$. We consider an iteration step

$$\gamma_{k+1} = \varphi_k(S_k)\,\gamma_k \ , \tag{44}$$

with φ_k to be chosen such that the following holds:

$$\exists_{c>0}[S_k \approx cI] \Rightarrow \|S_{k+1} - cI\| = O(\|S_k - cI\|^m) \ . \tag{45}$$

The condition (45) means that the asymptotic order of convergence of the iteration on the level of frame operators is equal to m.

So let $c > 0$ be such that $S_k \approx cI$. Note that

$$S_{k+1} = S_k \varphi_k^2(S_k) \tag{46}$$

by Theorem 1 in Sec. 3. Assuming the operators in (46) to be positive, so that square roots of them may be taken, we require that

$$S_k^{1/2} \varphi_k(S_k) = c^{1/2} I + O(\|S_k - cI\|^m) \ . \tag{47}$$

Up to errors $O(\|S_k - cI\|^m)$ we calculate

$$\begin{aligned} S_k^{-1/2} &= c^{-1/2}(I - (I - c^{-1}S_k))^{-1/2} \\ &= c^{-1/2} \sum_{i=0}^{m-1} (-1)^i \binom{-1/2}{i} (I - c^{-1}S_k)^i \\ &= c^{-1/2} \sum_{j=0}^{m-1} a_{mj}\, c^{-j}\, S_k^j \ . \end{aligned} \tag{48}$$

The quantities a_{mj} in (48) are given explicitly as

$$a_{mj} = \sum_{i=j}^{m-1} (-1)^{i+j} \binom{-1/2}{i} \binom{i}{i-j} , \qquad j = 0, ..., m-1 \ , \tag{49}$$

as can be obtained by binomial expansion of each of the terms $(I - c^{-1}S_k)^i$ in (48). This then suggests to take

$$\varphi_k(s) = \sum_{j=0}^{m-1} a_{mj}\, c^{-j}\, s^j \ . \tag{50}$$

The quantities c^{-j} are not known, however, and should be estimated one way or another. One can use, for instance

$$\frac{\|\gamma_k\|}{\|S_k^j \gamma_k\|} , \qquad j = 0, 1, ..., m-1 \ , \tag{51}$$

as an estimate for c^{-j}. Dividing through by the overall constant $\|\gamma_k\|$, we finally get the recursion step

$$\gamma_{k+1} = \sum_{j=0}^{m-1} a_{mj} \frac{S_k^j \gamma_k}{\|S_k^j \gamma_k\|} , \tag{52}$$

where the a_{mj} are given explicitly by (49). We may note here that $\sum_{j=0}^{m-1} a_{mj} = 1$ and that all terms $S_k^j\gamma_k/\|S_k^j\gamma_k\|$ at the right-hand side of (52) have unit norm.

We point out that we have argued heuristically here, our purpose only being to come up with a sound proposal for an iterative method with order of convergence m. In particular, the matter of taking square roots in (47) deserves special care, and actually nothing has been proved yet. In Sec. 6 and in [7], Sec. 8, these matters are taken care of in detail for the cases $m = 2$ and 3, respectively.

Examples: For $m = 2$ and $m = 3$ one gets the recursion steps

$$\gamma_{k+1} = \frac{3}{2}\,\frac{\gamma_k}{\|\gamma_k\|} - \frac{1}{2}\,\frac{S_k\gamma_k}{\|S_k\gamma_k\|} \tag{53}$$

and

$$\gamma_{k+1} = \frac{15}{8}\,\frac{\gamma_k}{\|\gamma_k\|} - \frac{5}{4}\,\frac{S_k\gamma_k}{\|S_k\gamma_k\|} + \frac{3}{8}\,\frac{S_k^2\gamma_k}{\|S_k^2\gamma_k\|}\,, \tag{54}$$

respectively.

5.2. *Iterations for g^d*

We next consider the problem of proposing m^{th} order convergent recursions for the approximation of g^d. To that end we consider algorithms of the type

$$\gamma_0 = g\,;\quad \gamma_{k+1} = \psi_k(S, S_k)\,\gamma_k\,, \qquad k = 0, 1, \ldots\,, \tag{55}$$

where S and S_k are the frame operators corresponding to (g, a, b) and (γ_k, a, b), respectively, and ψ_k is an analytic function of two variables. We first need to develop tools as in Sec. 3 for the present situation.

By induction it is seen from (55) that there are analytic functions f_k, g_k such that

$$\gamma_k = f_k(S)\,g\,, \qquad S_k = g_k(S)\,, \tag{56}$$

where $g_k(s) = s\,f_k^2(s)$, see Theorem 1 in Sec. 3. Assuming all operators to be positive it follows that

$$\gamma_k = S^{-1/2}\,S_k^{1/2}\,g\,, \qquad g^d = S^{-1}g = (SS_k)^{-1/2}\,\gamma_k\,. \tag{57}$$

For later use we also note that we have

$$(SS_k)^{1/2}\,\gamma_k = S_k\,g\,. \tag{58}$$

From the second item in (57) we see that

$$\left\| \frac{\gamma_k}{\|\gamma_k\|} - \frac{g^d}{\|g^d\|} \right\| = \left\| \frac{\gamma_k}{\|\gamma_k\|} - \frac{(SS_k)^{-1/2}\,\gamma_k}{\|(SS_k)^{-1/2}\,\gamma_k\|} \right\| , \tag{59}$$

and this shows that we are in a situation completely similar to what we have encountered in Sec. 3 for g^t. Accordingly, we should concentrate on convergence of $SS_k/\|SS_k\|$ to I.

We shall first show that $S_{k+1} = S_k\psi^2(S, S_k)$. Indeed, we have from (56) that

$$\gamma_{k+1} = \psi_k(S, g_k(S))\, f_k(S)\, g \ , \tag{60}$$

so that by Theorem 1 in Sec. 3.

$$S_{k+1} = S[\psi_k(S, g_k(S))\, f_k(S)]^2 = S f_k^2(S)\, \psi_k^2(S, g_k(S)) = S_k \psi_k^2(S, S_k) \ . \tag{61}$$

We next want ψ_k to satisfy the following:

$$\exists_{c>0}\,[SS_k \approx cI] \Rightarrow \|SS_{k+1} - cI\| = O(\|SS_k - cI\|^m) \ . \tag{62}$$

Here $m = 2, 3, \ldots$ is the envisaged convergence order of the method. Again assuming that all square roots may be taken, we let

$$Z_k = (SS_k)^{1/2} \ . \tag{63}$$

Then there holds $Z_{k+1} = Z_k\psi_k(S, S_k)$ by the above. Now assume that $c > 0$ is such that $Z_k \approx c^{1/2}I$. We require thus that

$$Z_k\psi_k(S, S_k) = c^{1/2}I + O(\|Z_k - c^{1/2}I\|^m) \ . \tag{64}$$

Up to errors $O(\|Z_k - c^{1/2}I\|^m)$ there holds, compare (48)

$$Z_k^{-1} = c^{-1/2}(I - (I - c^{-1/2}Z_k))^{-1} = c^{-1/2} \sum_{j=0}^{m-1} b_{mj}\, c^{-1/2j}\, Z_k^j \ , \tag{65}$$

where the b_{mj} have the explicit form

$$b_{mj} = (-1)^j \sum_{i=j}^{m-1} \binom{i}{i-j} \ , \qquad j = 0, \ldots, m-1 \ . \tag{66}$$

This suggests to take

$$\psi_k(S, S_k) = \sum_{j=0}^{m-1} b_{mj}\, c^{-j/2}\, Z_k^j \ , \tag{67}$$

so that the proposed recursion step becomes

$$\gamma_{k+1} = \sum_{j=0}^{m-1} b_{mj}\, c^{-j/2}\, Z_k^j\, \gamma_k \ . \tag{68}$$

There are now two problems with the proposal (68) that should be addressed. First, the numbers $c^{-j/2}$ at the right-hand side of (68) are unknown. We estimate these $c^{-j/2}$ as

$$\frac{\|\gamma_k\|}{\|Z_k^j \gamma_k\|} \ , \qquad j = 0, 1, ..., m-1 \ . \tag{69}$$

Secondly, we should find a way to compute $Z_k^j \gamma_k$ without having to take square roots of operators, see (63). For this we have (58) and accordingly for $i = 0, 1, ...$

$$Z_k^{2i} \gamma_k = (SS_k)^i \gamma_k \ ; \qquad Z_k^{2i+1} \gamma_k = (SS_k)^i\, S_k\, g \ . \tag{70}$$

Finally, dividing through by $\|\gamma_k\|$, we obtain as proposed recursion step

$$\gamma_{k+1} = \sum_{j=0}^{m-1} b_{mj}\, \frac{Z_k^j \gamma_k}{\|Z_k^j \gamma_k\|} \ , \tag{71}$$

where the b_{mj} are given in (66) and the $Z_k^j \gamma_k$ are to be computed according to (70). We note here that $\sum_{j=0}^{m-1} b_{mj} = 1$ and that all terms $Z_k^j \gamma_k / \|Z_k^j \gamma_k\|$ at the right-hand side of (71) have unit norm.

For use in Sec. 7 we mention the following result. When we have S, S_k, Z_k as above and $\gamma_{k+1} = \varphi_k(Z_k)\, \gamma_k$, with φ_k analytic around and positive on $\sigma(Z_k)$, then there holds

$$Z_{k+1} = Z_k\, \varphi_k(Z_k) \ . \tag{72}$$

Indeed, with (56) we see that

$$\gamma_{k+1} = \varphi_k((S\, g_k(S))^{1/2})\, f_k(S)\, g \tag{73}$$

so that by Theorem 1 in Sec. 3

$$S_{k+1} = S\, [\varphi_k((S\, g_k(S))^{1/2})\, f_k(S)]^2 = S\, f_k^2(S)\, \varphi_k^2(Z_k) = S_k \varpi_k^2(Z_k) \ . \tag{74}$$

Then multiplying by S and taking square roots in the far left- and the far right-hand side of (74) we get (72). In particular, we see in the case of (71) that

$$Z_{k+1} = \sum_{j=0}^{m-1} \frac{b_{mj}}{\|Z_k^j \gamma_k\|}\, Z_k^j \ . \tag{75}$$

As in Subsec. 5.1 we must point out that we haven't proved much in this subsection, our aim just being to develop a mechanism for soundly proposing recursion steps for the mth order convergent methods.

Examples: For $m = 2$ and 3 one gets the recursion steps

$$\gamma_{k+1} = 2\,\frac{\gamma_k}{\|\gamma_k\|} - \frac{S_k g}{\|S_k g\|} \tag{76}$$

and

$$\gamma_{k+1} = 3\,\frac{\gamma_k}{\|\gamma_k\|} - 3\,\frac{S_k g}{\|S_k g\|} + \frac{S\,S_k\,\gamma_k}{\|S\,S_k\,\gamma_k\|}\,, \tag{77}$$

respectively.

6. Analysis of Recursion II to Approximate g^t

In this section we consider the recursion

$$\gamma_0 = g\,;\quad \gamma_{k+1} = \tfrac{3}{2}\,\varepsilon_{k0}\gamma_k - \tfrac{1}{2}\,\varepsilon_{k1}\,S_k\,\gamma_k\,, \qquad k = 0, 1, \ldots\,, \tag{78}$$

with $\varepsilon_{k0} > 0$, $\varepsilon_{k1} > 0$, and we pay special attention to the case that

$$\varepsilon_{k0} = \|\gamma_k\|^{-1}\,, \qquad \varepsilon_{k1} = \|S_k\gamma_k\|^{-1}\,. \tag{79}$$

There holds now

$$\gamma_{k+1} = \varphi_k(S_k)\,\gamma_k\,; \qquad \varphi_k(s) = \tfrac{3}{2}\,\varepsilon_{k0} - \tfrac{1}{2}\,\varepsilon_{k1} s\,. \tag{80}$$

In order to apply Theorem 1 in Sec. 3 we need that $\varphi_k(s) > 0$, $s \in \sigma(S_k)$, and so we assume

$$\frac{\varepsilon_{k0}}{\varepsilon_{k1}} > \tfrac{1}{3}\,B_k\,. \tag{81}$$

We can now safely pass to square roots of frame operators, whence we let for $k = 0, 1, \ldots$

$$Z_k = S_k^{1/2}\,; \qquad E_k = A_k^{1/2} = \min\sigma(Z_k)\,,\quad F_k = B_k^{1/2} = \max\sigma(Z_k)\,. \tag{82}$$

Then there holds

$$Z_{k+1} = (S_k\varphi_k^2(S_k))^{1/2} = Z_k(\tfrac{3}{2}\,\varepsilon_{k0}I - \tfrac{1}{2}\,\varepsilon_{k1}Z_k^2)\,. \tag{83}$$

Since $\sigma(Z_k) \subset [E_k, F_k]$, we have by the spectral mapping theorem in a similar fashion as in (31)–(32) that

$$\frac{E_{k+1}}{F_{k+1}} \geq \frac{\min z(\frac{3}{2}\,\varepsilon_{k0} - \frac{1}{2}\,\varepsilon_{k1} z^2)}{\max z(\frac{3}{2}\,\varepsilon_{k0} - \frac{1}{2}\,\varepsilon_{k1} z^2)}\,, \tag{84}$$

where the min and max are taken over $[E_k, F_k]$.

In Fig. 2 we have plotted for fixed values of $\varepsilon_0 > 0$, $\varepsilon_1 > 0$ and of $E > 0$, $F > 0$ such that $\varepsilon_0/\varepsilon_1 > \frac{1}{3}\,F^2$ (see (81)) and $E < F$ the graph of the mapping

$$z > 0 \to z\,\varphi(z^2)\ ; \qquad \varphi(s) = \tfrac{3}{2}\,\varepsilon_0 - \tfrac{1}{2}\,\varepsilon_1 s, \tag{85}$$

and we consider the minimum and maximum of $z\,\varphi(z^2)$ over the interval $[E, F]$, where we distinguish between the cases

$$\text{(i) } E \leq \Big(\frac{\varepsilon_0}{\varepsilon_1}\Big)^{1/2} \leq F\,, \quad \text{(ii) } E \leq F < \Big(\frac{\varepsilon_0}{\varepsilon_1}\Big)^{1/2}, \quad \text{(iii) } \Big(\frac{\varepsilon_0}{\varepsilon_1}\Big)^{1/2} < E \leq F\,. \tag{86}$$

It is thus seen that

$$\text{Case (i)}: \quad \frac{\text{min over } [E,F]}{\text{max over } [E,F]} = \frac{\text{value at } (\varepsilon_0/\varepsilon_1)^{1/2}}{\text{max over } \{E,F\}}\,, \tag{87}$$

$$\text{Case (ii)}: \quad \frac{\text{min over } [E,F]}{\text{max over } [E,F]} = \frac{\text{value at } E}{\text{value at } F}\,, \tag{88}$$

$$\text{Case (iii)}: \quad \frac{\text{min over } [E,F]}{\text{max over } [E,F]} = \frac{\text{value at } F}{\text{value at } E}\,, \tag{89}$$

where the minima, maxima and values refer to the function $z\,\varphi(z^2)$ in (85). Note that the condition $\varepsilon_0/\varepsilon_1 > \frac{1}{3}\,F^2$ ensures that $z\,\varphi(z^2) > 0$ for $z \in [E, F]$.

Returning to (84), an elementary analysis (detailed in [7], Sec. 4) shows that in case (i), with ε_{0k}, ε_{1k}, E_k, F_k instead of ε_0, ε_1, E, F, there holds

$$\frac{E_{k+1}}{F_{k+1}} \geq \frac{F_k}{E_k}\left(\frac{3}{2} - \frac{1}{2}\Big(\frac{F_k}{E_k}\Big)^2\right), \tag{90}$$

with equality when $E_k = (\varepsilon_{0k}/\varepsilon_{1k})^{1/2}$. The right-hand side of (87) is increasing in $E_k/F_k \in (0,1]$ and exceeds E_k/F_k if and only if $E_k/F_k > \frac{1}{2}\sqrt{2}$.

Hence in case (i) there holds

$$\frac{E_k}{F_k} > \tfrac{1}{2}\sqrt{2} \Rightarrow \frac{E_{k+1}}{F_{k+1}} > \frac{E_k}{F_k} \,. \tag{91}$$

A further elementary analysis shows that, in case (i), the maximum value of the right-hand side of (86) equals

$$\tfrac{1}{2}\,(E_k F_k^2 + E_k^2 F_k)/(\tfrac{1}{3}\,(E_k^2 + E_k F_k + F_k^2))^{3/2} \,, \tag{92}$$

and occurs when

$$\frac{\varepsilon_{0k}}{\varepsilon_{1k}} = \tfrac{1}{3}\,(E_k^2 + E_k F_k + F_k^2) \,. \tag{93}$$

In terms of the frame operators $S_k = Z_k^2$ and their frame bounds $A_k = E_k^2$, $B_k = F_k^2$ with frame bound quotients $Q_k = A_k/B_k$ the results for case (i) can be summarized as follows. We are in case (i) if and only if

$$\frac{\varepsilon_{0k}}{\varepsilon_{1k}} \in [A_k, B_k] \,, \tag{94}$$

and this condition can be satisfied in various ways, see Theorem 4 in Sec. 3, for instance by taking ε_{0k}, ε_{1k} as in (79). Also, in case (i), there holds

$$Q_{k+1} \geq Q_k^{-1}(\tfrac{3}{2} - \tfrac{1}{2}\,Q_k^{-1}) = 1 - \frac{Q_k - \frac{1}{4}}{Q_k^2}\,(1 - Q_k)^2 \,, \tag{95}$$

and the right-hand side of (95) exceeds Q_k when $Q_k > \frac{1}{2}$. It is thus seen that $Q_k \to 1$, at least quadratically and monotonically, when $Q_0 = A/B > \frac{1}{2}$, where A and B are the best frame bounds of (g, a, b). Finally, still in case (i), when $\sigma(S) = [A, B]$ so that $\sigma(S_k) = [A_k, B_k]$ for all $k = 0, 1, \ldots$, we have from (92) that

$$Q_{k+1} \leq \frac{27}{4}\,\frac{(Q_k^{1/2} + Q_k)^2}{(1 + Q_k^{1/2} + Q_k)^3} =$$

$$= 1 - \frac{(2 + Q_k^{1/2})^2\,(\frac{1}{2} + Q_k^{1/2})^2}{(1 + Q_k^{1/2})^2\,(1 + Q_k^{1/2} + Q_k)^3}\,(1 - Q_k)^2 \,, \tag{96}$$

with equality if and only if

$$\frac{\varepsilon_{0k}}{\varepsilon_{1k}} = \tfrac{1}{3}\,(A_k + (A_k B_k)^{1/2} + B_k) \,. \tag{97}$$

Hence, the worst-case tightening up result as given in (95) has a right-hand side $\approx 1 - \frac{3}{4}\,(1 - Q_k)^2$ and the best-case tightening up result in (96) has a right-hand side $\approx 1 - \frac{3}{16}\,(1 - Q_k)^2$ when Q_k is close to 1. This latter case is often more realistic, especially when we manage to take $\varepsilon_{0k}/\varepsilon_{1k}$ well in the middle of the interval $[A_k, B_k]$. We may also observe that the quantity at the right-hand side of (96) exceeds Q_k, no matter how small Q_k is. We conclude from this discussion that the sufficient condition $Q_0 = A/B > \frac{1}{2}$ for monotone and quadratic convergence with the choice ε_{0k}, ε_{1k} as in (79) is quite often much too stringent.

We now discuss the other two cases in (86) and we start with case (ii). It follows from an elementary analysis, or an inspection of the graph of $z\,\varphi(z^2)$, see Fig. 2(ii) and (85), that in case (ii) one has

$$\frac{E_{k+1}}{F_{k+1}} > \frac{E_k}{F_k} , \tag{98}$$

no matter how small E_k/F_k is. Hence this is a safe mode for the iteration (78) in which in each step the frame gets tighter. One can get in this safe mode as follows. Replace

$$g \text{ by } g/\hat{B}^{1/2} , \qquad S \text{ by } S/\hat{B} , \tag{99}$$

where $\hat{B}$ is any number $\geq B$ (the best upper frame bound of (g, a, b)), and take $\varepsilon_{k0} = \varepsilon_{k1} = 1$. It is easy to see that, directly in terms of S_k and A_k, B_k, there holds

$$A_{k+1} = A_k(\tfrac{3}{2} - \tfrac{1}{2}\,A_k^2) = 1 - \tfrac{1}{2}\,(A_k + 2)(1 - A_k)^2 , \tag{100}$$

$$B_{k+1} = B_k(\tfrac{3}{2} - \tfrac{1}{2}\,B_k^2) = 1 - \tfrac{1}{2}\,(B_k + 2)(1 - B_k)^2 , \tag{101}$$

(equality here since $A_k, B_k \in \sigma(S_k)$ being best frame bounds). From this it follows that both A_k and B_k tend to 1, monotonically and quadratically. Hence, in this safe mode (ii), there is quadratic convergence no matter how small the frame bound ratio A/B of the initial frame (g, a, b) is. Of course, the constants governing the convergence in the safe mode are not as good as those in the smart but risky mode (i). And also, when both A and B are very small compared to 1, it may take a large number of iterations before actual (quadratic) convergence takes place.

Case (ii) in (86) is neither smart nor safe and should therefore be avoided at all times.

We conclude this section by pointing at the opportunity of switching between the safe mode (ii) and the smart mode (i): we have $\gamma_k^t = g^t$, no matter how we choose ε_{k0}, ε_{k1} and whether we prescale g and S as in (99) or not. This means that one can combine the advantages of both modes by first scaling g and S as in (99) and iterating with $\varepsilon_{k0} = \varepsilon_{k1} = 1$ until one is confident that $A_k/B_k > \frac{1}{2}$ and then switching to mode (i) with ε_{k0}, ε_{k1} as in (79), for instance.

7. Analysis of Recursion IV to Approximate g^d

In this section we consider the recursion

$$\gamma_0 = g \; ; \qquad \gamma_{k+1} = 2\delta_{k0}\gamma_k - \delta_{k1}\, S_k\, g \,, \quad k = 0, 1, \ldots \,, \tag{102}$$

with $\delta_{k0} > 0$, $\delta_{k1} > 0$, and we pay special attention to the case that

$$\delta_{k0} = \|\gamma_k\|^{-1} \,, \qquad \delta_{k1} = \|S_k\, g\|^{-1} \,. \tag{103}$$

We follow the plan of the analysis of recursion II as given in Sec. 6 rather closely.

We have, see for instance (61), (72)

$$S_{k+1} = S_k(2\delta_{k0}I - \delta_{k1}(SS_k)^{1/2})^2 \,, \tag{104}$$

provided that S_k is a positive operator so that the square root at the right-hand side can be taken. Hence, with

$$Z_k = (SS_k)^{1/2} \; ; \qquad E_k = \min \sigma(Z_k) \,, \quad F_k = \max \sigma(Z_k) \,, \tag{105}$$

we assume that

$$\frac{\delta_{k0}}{\delta_{k1}} > \tfrac{1}{2}\, F_k \,, \tag{106}$$

so that the operator at the right-hand side of (104) is positive indeed. Multiplying either side of (104) by S and taking square roots, we get (also see (72))

$$Z_{k+1} = Z_k(2\delta_{k0}I - \delta_{k1}Z_k) \,. \tag{107}$$

We are now practically in a similar position as in Sec. 6, (82)–(84). Accordingly, we get

$$\frac{E_{k+1}}{F_{k+1}} \geq \frac{\min z(2\delta_{k0} - \delta_{k1}z)}{\max z(2\delta_{k0} - \delta_{k1}z)} \,, \tag{108}$$

where min and max are taken over $[E_k, F_k]$, and three cases

$$\text{(i) } E_k \leq \frac{\delta_{0k}}{\delta_{1k}} \leq F_k \,, \qquad \text{(ii) } E_k \leq F_k < \frac{\delta_{0k}}{\delta_{1k}} \,, \qquad \text{(iii) } \frac{\delta_{0k}}{\delta_{1k}} < E_k \leq F_k \tag{109}$$

have to be considered. Here case (i) corresponds to the smart but risky mode, case (ii) corresponds to the safe but conservative mode, while case (iii) is the mode that is neither smart nor safe.

The results are then as follows. Let $Q_k = E_k/F_k$. In case (i) in (109) we have

$$Q_{k+1} \geq Q_k^{-1}(2 - Q_k^{-1}) = 1 - \left(\frac{1 - Q_k}{Q_k}\right)^2 \,. \tag{110}$$

The right-hand side of (110) is increasing in $Q_k \in (0,1]$ and exceeds Q_k when $Q_k > \frac{1}{2}(\sqrt{5}-1)$. Therefore we have that $Q_k \to 1$, monotonically and at least quadratically, when $Q_0 = E_0/F_0 = A/B > \frac{1}{2}(\sqrt{5}-1)$, provided that we are in case (i) during the iteration process. This latter condition can be satisfied in many ways, see Theorem 4 in Sec. 3, for instance, by taking δ_{k0}, δ_{k1} as in (103). Indeed, for this choice of δ_{k0}, δ_{k1}, we have, see (58),

$$\frac{\delta_{0k}}{\delta_{1k}} = \frac{\|S_k g\|}{\|\gamma_k\|} = \frac{\|Z_k \gamma_k\|}{\|\gamma_k\|} \in [E_k, F_k] \ . \tag{111}$$

As a best-case result in case (i) we have the following. Assume that $\sigma(S) = [A,B]$, so that $\sigma(Z_k) = [A_k, B_k]$ for all k. Then there holds

$$Q_{k+1} \le \frac{4Q_k}{(1+Q_k)^2} = 1 - \left(\frac{1-Q_k}{1+Q_k}\right)^2 , \tag{112}$$

with equality if and only if

$$\frac{\delta_{0k}}{\delta_{1k}} = \tfrac{1}{2}(E_k + F_k) \ . \tag{113}$$

Hence, the worst-case tightening up result (110) can be replaced by the best-case tightening up result (112) when we succeed in choosing δ_{0k}/δ_{1k} right in the middle of $[E_k, F_k]$. Note also that the right-hand side of (112) exceeds Q_k, no matter how small Q_k is.

In the safe mode (ii) in (109) we have

$$\frac{E_{k+1}}{F_{k+1}} > \frac{E_k}{F_k} \ . \tag{114}$$

This safe mode is reached by replacing g and S as in (99) and by taking $\delta_{k0} = \delta_{k1} = 1$. Then we have $0 < E_k \le F_k < 1$, and

$$E_{k+1} = E_k(2-E_k) = 1-(1-E_k)^2 \ , \qquad F_{k+1} = F_k(2-F_k) = 1-(1-F_k)^2 \tag{115}$$

(equality here since $E_k, F_k \in \sigma(Z_k)$). Hence in this safe mode there is quadratic convergence, no matter how small the initial frame bound ratio A/B is. However, the convergence constants are not as good as in the smart mode, and when A and B are very small it can take many iterations before actual (quadratic) convergence takes place.

8. Summary of Results for Iterations III and V

In this section we consider the iterations III and V for which the recursion steps are given by

$$\gamma_{k+1} = \tfrac{15}{8}\,\varepsilon_{k0}\gamma_k - \tfrac{5}{4}\,\varepsilon_{k1}\, S_k\, \gamma_k + \tfrac{3}{8}\,\varepsilon_{k2}\, S_k^2\, \gamma_k \ , \tag{116}$$

and

$$\gamma_{k+1} = 3\delta_{k0}\gamma_k - 3\delta_{k1}\, S_k\, g + \delta_{k2}\, SS_k\, \gamma_k$$

$$= 3\delta_{k0}\gamma_k - 3\delta_{k1}\, Z_k\, \gamma_k + \delta_{k2}\, Z_k^2\, \gamma_k \tag{117}$$

(with $Z_k = (SS_k)^{1/2}$ in (117)), for the approximation of g^t and g^d, respectively. These recursions are considered in all detail in [7], Sec. 8 and 9, with particular attention for smart modes. We summarize the results of [7] here and present some supplements regarding safe modes.

The recursions (116) and (117) are analyzed under conditions that arise when one chooses

$$\varepsilon_{ki} = \|S_k^i \gamma_k\|^{-1} \quad \text{and} \quad \delta_{ki} = \|Z_k^i \gamma_k\|^{-1} \ , \tag{118}$$

respectively. In the case of recursion (116), with frame operators S_k of (γ_k, a, b) having frame bounds A_k and B_k, there are the restrictions

$$A_k \le u_k := \frac{\varepsilon_{k1}}{\varepsilon_{k2}} \le B_k \ , \qquad \frac{2A_k B_k}{A_k^2 + B_k^2} \le v_k := \frac{\varepsilon_{k0}\varepsilon_{k2}}{\varepsilon_{k1}^2} \le 1 \ . \tag{119}$$

The two inequalities in (119) correspond to the inequalities

$$C \le \frac{\|T^2 f\|}{\|Tf\|} \le D \ , \qquad \frac{2CD}{C^2 + D^2} \le \frac{\|Tf\|^2}{\|f\|\,\|T^2 f\|} \le 1 \ , \tag{120}$$

valid for a positive linear operator T of a Hilbert space H and $f \in H$, also see (22), with $C = \min \sigma(T)$, $D = \max \sigma(T)$. The restrictions on δ_{ki} in (117) are similar, except that now in (119) we have $u_k = \delta_{k1}/\delta_{k2}$, $v_k = \delta_{k0}\delta_{k2}/\delta_{k1}^2$ and $E_k = \min Z_k$, $F_k = \max Z_k$ instead of A_k, B_k.

Actually, the recursions (116) and (117) are analyzed under a slightly stronger restriction than what the Kantorovich inequality in the second item in (119) gives. This stronger version of the second item in (119) reads

$$\frac{u_k^{-1}\, A_k\, B_k}{(A_k^2 + B_k^2 - u_k^2)^{1/2}} \le v_k \le 1 \ . \tag{121}$$

This sharpening is relevant in the present context in the following sense. In [7] it is shown that for any four numbers A, B, u, t with

$$0 < A \le \frac{AB}{(A^2 + B^2 - u^2)^{1/2}} \le t \le u \le B < \infty \ , \tag{122}$$

there is a Gabor frame $(g, a = 1, b = 1)$ with frame operator S having best frame bounds A, B and such that

$$\frac{\|Sg\|}{\|g\|} = t \ , \qquad \frac{\|S^2 g\|}{\|Sg\|} = u \ . \tag{123}$$

Under condition (119), strengthened according to (121), it is shown in [7] for recursion (116) that

$$\frac{A_k}{B_k} > \frac{3}{7} \Rightarrow \frac{A_{k+1}}{B_{k+1}} > \frac{A_k}{B_k} \;, \tag{124}$$

and that cubic and monotone convergence occurs when $A/B > 3/7$. Under the analog of condition (119) for recursion (117), strengthened according to (121), it is shown in [7] that

$$\frac{E_k}{F_k} > Q \Rightarrow \frac{E_{k+1}}{F_{k+1}} > \frac{E_k}{F_k} \;, \tag{125}$$

and that cubic and monotone convergence occurs when $A/B > Q$. Here

$$Q = 0.513829766... = \frac{3}{2W^3} - \left(\left(\frac{3}{2W^3}\right)^2 - 1\right)^{1/2} \tag{126}$$

with W the unique solution in $(0, (3/2)^{1/2})$ of

$$9W^2 - 16W^3 + 9W^4 + 6W^5 + 6W^7 + 2W^9 = 24 \;. \tag{127}$$

No effort is spent in [7] to find the best cases of the recursions (116), (117) under the condition that $\sigma(S) = [A, B]$.

Safe but conservative versions of the recursions (116), (117) can be obtained by replacing g and S as in (99) and by taking

$$\varepsilon_{ki} = 1 \,, \quad \delta_{ki} = 1 \;, \qquad k = 0, 1, \ldots, \quad i = 0, 1, 2 \;. \tag{128}$$

In that case there holds

$$0 < A_k \le B_k < 1 \,, \quad 0 < E_k \le F_k < 1 \;, \qquad k = 0, 1, \ldots \;. \tag{129}$$

Furthermore, the quantities $R_k := A_k^{1/2}, B_k^{1/2}$ are recursively given as $R_0 = A^{1/2}, B^{1/2}$ and

$$R_{k+1} = R_k(\tfrac{15}{8} - \tfrac{5}{4}\,R_k^2 + \tfrac{3}{8}\,R_k^4) = 1 - (\tfrac{3}{8}\,R_k^2 + \tfrac{9}{8}\,R_k + 1)(1 - R_k)^3 \;,$$

$$k = 0, 1, \ldots \;, \tag{130}$$

and the quantities $U_k := E_k, F_k$ are recursively given as $U_0 = A, B$ and

$$U_{k+1} = U_k(3 - 3U_k + U_k^2) = 1 - (1 - U_k)^3 \;, \qquad k = 0, 1, \ldots \;. \tag{131}$$

It follows then that both A_k, B_k and E_k, F_k converge cubically and monotonically to 1, no matter how small A_k/B_k or E_k/F_k are.

9. Concluding Remarks

We have presented three iterative algorithms for the approximation of the canonical tight window $g^t = S^{-1/2}g$ associated with a Gabor frame (g, a, b) with frame operator S and two iterative algorithms for the approximation of the canonical dual window $g^d = S^{-1}g$ associated with (g, a, b). These algorithms require in the kth recursion step the application of the current frame operator S_k (and, in one instance of the algorithms for g^t, of S_k^{-1}) to the current window γ_k and/or to g for algorithms with envisaged quadratic convergence. For the algorithms with envisaged cubic convergence, the operators S_k^2 or SS_k have to be applied, in addition, to γ_k or g. We have developed a number of tools to analyze these algorithms, where a key role is played by the spectral mapping theorem, basic frame operator calculus and basic inequalities in Hilbert space operator theory such as the Kantorovich inequality. We have demonstrated the effectiveness of the developed tools by redoing the analysis in [6] for the approximation of g^t using frame operator inversions. We have presented a rationale for proposing iterative algorithms of the described type with an envisaged convergence order $m = 2, 3, \ldots$, and we have given a detailed analysis of the algorithms for g^t and g^d with quadratic convergence ($m = 2$). Furthermore, we have summarized the results of [7], Secs. 8–9, in which detailed analyses are given of the algorithms with cubic convergence ($m = 3$).

All algorithms we have considered have smart versions, in which excellent convergence behavior is exhibited conditionally on the condition number of the initial frame operator S, and safe versions, in which there is unconditional convergence of the required order with suboptimal convergence constants. One can freely switch between the safe version and the smart version without altering the limiting window g^t or g^d. Preliminary experiments, with the standard Gaussian window $g(t) = 2^{1/4}\exp(-\pi t^2)$ and $a = b = 1/\sqrt{2}$, have shown that the smart versions produce 10^{-15} accurate approximations within 4–7 steps for the algorithms with quadratic convergence and within 2 or 3 steps for the algorithms with cubic convergence. The number of steps required for the safe versions is typically a factor $1\frac{1}{2}$–2 larger.

Acknowledgments

The author wishes to thank Hans Feichtinger and Thomas Strohmer for many fruitful discussions during the last few years on this subject as well as for doing preliminary experiments with various algorithms. Thanks are also due to Mario Hampejs, whose observations in [5] led the author to the introduction of the safe modes of the algorithms, and to Mark Lammers for sending [9] prior to publication. The work was partially done while the author was visiting the Institute for Mathematical Sciences, National University of Singapore in September 2003; the visit was supported by the Institute.

References

1. I. Daubechies, Ten Lectures on Wavelets, SIAM, 1992.
2. H. G. Feichtinger and T. Strohmer, Gabor Analysis and Algorithms: Theory and Applications, Birkhäuser, 1998.
3. H. G. Feichtinger, F. Luef and T. Werther, A guided tour from linear algebra to the foundations of Gabor analysis, Gabor and Wavelet Frames, World Scientific, 2007.
4. K. Gröchenig, Foundations of Time-Frequency Analysis, Birkhäuser, 2001.
5. M. Hampejs, Computing the dual and the tight Gabor atom, private communication, 2002.
6. A. J. E. M. Janssen and T. Strohmer, Characterization and computation of canonical tight windows for Gabor frames, J. Fourier Anal. Appl. 8 (1) (2002), 1–28.
7. A. J. E. M. Janssen, Analysis of some fast algorithms to compute canonical windows for Gabor frames, preprint, 2002.
8. L. V. Kantorovich, Functional Analysis and Applied Mathematics, U.S. Department of Commerce, National Bureau of Standards, NBS Rep. 1509, 1952.
9. M. C. Lammers, Convolution for Gabor systems and Newton's method, private communication, 2003.

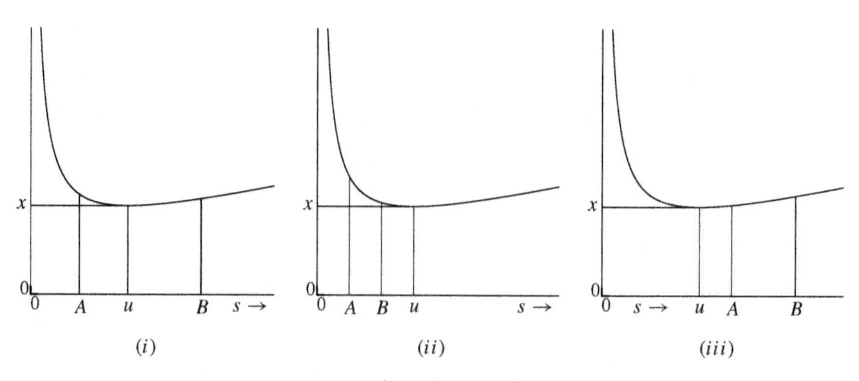

Fig. 1. Graph of the mapping $s \to s(\frac{1}{2}\,\alpha + \frac{1}{2}\,\beta\, s^{-1})^2$ with $\alpha > 0$, $\beta > 0$ and $0 < A < B$ such that (i) $A \le u \le B$, (ii) $A \le B < u$, (iii) $u < A \le B$. Here $u = \beta/\alpha$ and $x = \alpha\beta$ are both taken 1.

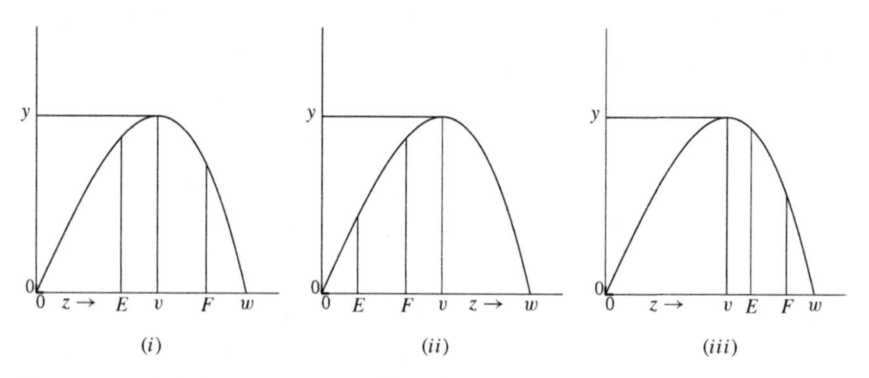

Fig. 2. Graph of the mapping $z \to z(\frac{3}{2}\,\varepsilon_0 - \frac{1}{2}\,\varepsilon_1 z^2)$ with $\varepsilon_0 > 0$, $\varepsilon_0 > 0$ and $0 < E < F$ such that (i) $E \le v \le F$, (ii) $E \le F < v$, (iii) $v < E \le F$. Here $v = (\varepsilon_0/\varepsilon_1)^{1/2}$ and $y = (\varepsilon_0^3/\varepsilon_1)^{1/2}$ are both taken 1, and $w = (3\varepsilon_0/\varepsilon_1)^{1/2}$.

GABOR ANALYSIS, NONCOMMUTATIVE TORI AND FEICHTINGER'S ALGEBRA

Franz Luef

NuHAG, Faculty of Mathematics, University of Vienna
Nordbergstrasse 15, A-1090 Vienna, Austria
E-mail: franz.luef@univie.ac.at

"Often in mathematics, understanding comes from generalization, instead of considering the object per se when one tries to find the concepts which embody the power of the object."

Alain Connes

We point out a connection between Gabor analysis and noncommutative analysis. Especially, the strong Morita equivalence of noncommutative tori appears as underlying setting for Gabor analysis, since the construction of equivalence bimodules for noncommutative tori has a natural formulation in the notions of Gabor analysis. As an application we show that Feichtinger's algebra is such an equivalence bimodule. Furthermore, we present Connes's construction of projective modules for noncommutative tori and the relevance of a generalization of Wiener's lemma for twisted convolution by Gröchenig and Leinert. Finally we indicate an approach to the biorthogonality relation of Wexler-Raz on the existence of dual atoms of a Gabor frame operator based on results about Morita equivalence.

1. Introduction

By definition, Morita equivalent algebras $\mathcal{A}$ and $\mathcal{B}$ have equivalent categories of (left) modules. The main theorem of algebraic Morita theory states that Morita equivalences are implemented by a $(\mathcal{A},\mathcal{B})$-*bimodule* X.

In [24] Morita proved the fundamental theorem, showing that this bimodule is invertible if and only if the bimodule is projective and finitely generated as a left $\mathcal{A}$-module and as a right $\mathcal{B}$-module and $\mathcal{A} \to End_{\mathcal{B}}(X)$ and $\mathcal{B} \to End_{\mathcal{A}}(X)$ are algebra isomorphisms. For further discussion on the Morita equivalence of algebras and the relevant modules and functors we refer to the recent paper [2].

In the seminal papers [29,30,31] Rieffel developed the notion of *strong Morita equivalence*, which is an extension of Morita equivalence of algebras to the setting of C^*-algebras. The relevant category of modules over a C^*-algebra, to be preserved under strong Morita equivalence, is that of Hilbert spaces on which the C^*-algebra acts through bounded operators. More precisely, for a given C^*-algebra $\mathcal{A}$, we consider the category $Herm(\mathcal{A})$ whose objects are pairs $(\mathcal{H}, \rho)$, where $\mathcal{H}$ is a Hilbert space and $\rho : \mathcal{A} \to \mathcal{B}(\mathcal{H})$ is a nondegenerate $*$-homomorphism of algebras, and morphisms are bounded intertwiners. Since we are dealing with more elaborate modules, it is natural that a bimodule giving rise to a functor $Herm(\mathcal{B}) \to Herm(\mathcal{A})$ should be equipped with extra structure. If $(\mathcal{H}, \rho) \in Herm(\mathcal{B})$ and ${}_{\mathcal{A}}V_{\mathcal{B}}$ is an $(\mathcal{A}, \mathcal{B})$-bimodule, then V itself is equipped with an inner product $\langle ., . \rangle_{\mathcal{B}}$ with values in $\mathcal{B}$. More precisely, let V be a right $\mathcal{B}$-module. Then a $\mathcal{B}$-valued inner product $\langle ., . \rangle_{\mathcal{B}}$ on V is a $\mathbb{C}$-sesquilinear pairing $V \times V \to \mathcal{B}$ (linear in the second argument) such that, for all $f_1, f_2 \in V$ and $T \in \mathcal{B}$, we have

(1) $\langle f_1, f_2 \rangle_{\mathcal{B}} = \langle f_2, f_1 \rangle_{\mathcal{B}}^*$
(2) $\langle f_1, f_2 T \rangle_{\mathcal{B}} = \langle f_1, f_2 \rangle_{\mathcal{B}} T$
(3) $\langle f_1, f_1 \rangle_{\mathcal{B}} > 0$ if $f_1 \neq 0$.

Inner products on left modules are defined analogously, but linearity is required in the first argument. One can show that $\|f\|_{\mathcal{B}} := \|\langle f, f \rangle_{\mathcal{B}}\|^{1/2}$ is a norm in V. A (right) **Hilbert $\mathcal{B}$-module** is a (right) $\mathcal{B}$-module V together with a $\mathcal{B}$-valued inner product $\langle ., . \rangle_{\mathcal{B}}$ so that V is complete with respect to $\|.\|_{\mathcal{B}}$. Our treatment of strong Morita equivalence is largely based on a recent paper of Bursztyn and Weinstein about the connection of Poisson geometry and noncommutative geometry, [2].

In the study of wavelets Rieffel/Packer and Wood have recently used Hilbert C^*-modules, see [25,26,42]. In the context of Gabor analysis we want to mention the work of Casazza/Lammers [3], but their work is unrelated to our investigations.

The main goal of this paper is that Rieffel's results on the strong Morita equivalence of noncommutative tori have a natural interpretation in terms of Gabor analysis. Recall that

$$T_x f(t) = f(t-x)$$

denotes a *translation* by $x \in \mathbb{R}^d$ for f in $L^2(\mathbb{R}^d)$ and

$$M_\omega f(t) = e^{2\pi i\omega t} f(t)$$

is a *modulation* by $\omega \in \widehat{\mathbb{R}}^d$. More precisely, we show – following Rieffel [35] – that the C^*-algebra $C^*(\Lambda)$ generated by time-frequency shifts $\pi(\lambda) = M_\omega T_x$ for $\lambda = (x,\omega)$ in a lattice Λ of the time-frequency plane $\mathbb{R}^d \times \widehat{\mathbb{R}}^d$ is Morita equivalent to the C^*-algebra of time-frequency shifts $C^*(\Lambda^0)$ generated by $\pi(\lambda^0)$ for λ^0 in the *adjoint lattice* Λ^0, see Section 2 for the definition of the adjoint lattice.

Our main result is that *Feichtinger's algebra* $S_0(\mathbb{R}^d)$ is a bimodule for the C^*-algebras $(C^*(\Lambda), C^*(\Lambda^0))$ or an *equivalence* bimodule as called by Rieffel.

Rieffel used Schwartz's space $\mathcal{S}(\mathbb{R}^d)$ in his construction of a bimodule for the C^*-algebras $(C^*(\Lambda), C^*(\Lambda^0))$. Therefore, our investigations are a further realization of a general strategy of Feichtinger, that consists in using $S_0(\mathbb{R}^d)$ is a good substitute for $\mathcal{S}(\mathbb{R}^d)$ in many situations.

As examples we state Feichtinger's seminal paper [8], Reiter's work on Weil's construction of the metaplectic group [28] and Gröchenig/ Leinert's result on the irrational case conjecture in Gabor analysis [19]. The present paper may be seen as a companion to [19], which provides some reasons for the relevance of noncommutative tori in Gabor analysis. On the other hand we show that Rieffel obtained the Janssen representation of Gabor frame operators, and the Fundamental Identity of Gabor analysis for functions in the Schwartz-Bruhat space $\mathcal{S}(G)$ for a locally compact abelian group G and a closed subgroup D in $G \times \widehat{G}$ already in 1988. More concretely, Rieffel observed that

$$\langle f, g\rangle_\Lambda = \sum_{\lambda\in\Lambda} \langle f, \pi(\lambda)g\rangle \pi(\lambda) \tag{1}$$

is a $C^*(\Lambda)$-valued inner product for f, g in $\mathcal{S}(\mathbb{R}^d)$.

It comes quite unexpectedly that Rieffel's work on Morita equivalence for noncommutative tori deals with the same mathematical structure as

Janssen's work on Gabor frame operators, [20]. The main goal of this paper is an exploration of this observation.

The paper is organized as follows: In Section 2 we discuss operator algebras of time-frequency shifts for a separated set and we show that there is a correspondence between symmetry of the separated set in $\mathbb{R}^d \times \widehat{\mathbb{R}}^d$ and the structure of the corresponding operator algebra. In this context we review some well-known results on projective representations and we define the adjoint set of an arbitrary separated set in $\mathbb{R}^d \times \widehat{\mathbb{R}}^d$ and discuss its relation to the structure of the commutant of the operator algebra.

In Section 3 we define noncommutative tori and Feichtinger's algebra. In this context we show that (1) is a Hilbert $C^*(\Lambda)$-valued inner product for f, g in $S_0(\mathbb{R}^d)$ using functorial properties of $S_0(\mathbb{R}^d)$. Furthermore we introduce the Short-Time Fourier Transform and its properties for functions in Feichtinger's algebra $S_0(\mathbb{R}^d)$.

In Section 4 we prove our main theorem, by showing that Feichtinger's algebra is a bimodule with respect to the pair of C^*-algebras $C^*(\Lambda)$ and $C^*(\Lambda^0)$. We discuss the notion of strong Morita equivalence in this setting and its relation to Gabor expansions, Gabor frame operators, Λ-invariant operators for a lattice Λ in $\mathbb{R}^d \times \widehat{\mathbb{R}}^d$. In the proof of Rieffel's associativity condition we observe its equivalence to the Fundamental Identity of Gabor Analysis and the existence of Janssen representation for Gabor frame operators, [35]. Our presentation follows [35], where he used the Poisson's summation formula for symplectic Fourier transform with respect to a lattice and its adjoint lattice in the proof of the Fundamental Identity of Gabor Analysis. Furthermore, we present the Gröchenig/Leinert [19] result on Wiener's lemma for twisted convolution in the context of Connes' construction of projective modules over $C^*(\Lambda)$.

In Section 5 we apply our results on strong Morita equivalence of the C^*-algebras $(C^*(\Lambda), C^*(\Lambda^0))$ to the structure of Gabor frames. Especially we derive the Wexler-Raz biorthogonality principle for the existence of dual windows for Gabor frames from the relation between traces on the Morita equivalent C^*-algebras on $C^*(\Lambda)$ and $C^*(\Lambda^0)$, respectively.

2. Operator Algebras of Time-Frequency Shifts

The most general framework in Gabor analysis builds reconstruction formulas for functions on $\mathbb{R}^d$ from a set of time-frequency shifts $\mathcal{G} = \{\pi(X_j) : X_j \in A\}$ for a countable subset $A = \{X_j = (x_j, \omega_j) : j \in J\}$ with $\inf_{j,k} |X_j - X_k| > \delta > 0$ in the time-frequency plane $\mathbb{R}^d \times \widehat{\mathbb{R}}^d$. We will refer to this setting as *non-uniform Gabor systems.*

First we derive some results about the structure of the operator algebra of a non-uniform Gabor system $\mathcal{G}$ from general principles. First the *commutant* $\mathcal{G}'$ of all bounded operators on $L^2(\mathbb{R}^d)$ commuting with every operator in $\mathcal{A}$ is a unital Banach algebra of time-frequency shifts with respect to operator composition and the operator norm, because $\mathcal{A}$ is a subset of unitary (bounded) operators on $L^2(\mathbb{R}^d)$. Furthermore, if $\mathcal{G}$ is generated by a set A which is symmetric with respect to the origin, then $\mathcal{G}$ is invariant under taking adjoints, moreover $\mathcal{G}$ is actually a group of time-frequency shifts acting on $L^2(\mathbb{R}^d)$. We have the following chains between $\mathcal{G}$ and its nth commutant $\mathcal{G}^{(n)}$:

$$\begin{aligned} \mathcal{G} \subset \mathcal{G}^{''} = \mathcal{G}^{(4)} = \cdots = \\ \mathcal{G}^{'} = \mathcal{G}^{(3)} = \mathcal{G}^{(5)} = \cdots . \end{aligned}$$

From now on, let our set of time-frequency shifts A be a lattice Λ of $\mathbb{R}^d \times \widehat{\mathbb{R}}^d$ then $\mathcal{G} = \mathcal{G}^{''}$. The associated Gabor system is called to be *regular.* These observations indicate a strong relationship between the symmetry of the set A and the structure of the corresponding operator algebra $\mathcal{G}$ and its commutant $\mathcal{G}^{'}$. In the following we discuss the connection between the set A of points in $\mathbb{R}^d \times \widehat{\mathbb{R}}^d$ and the set A^0 of points, which generate the commutant $\mathcal{G}^{'}$. This investigation yields a beautiful structure of regular Gabor systems, which is the very reason for the existence of all of our subsequent results. Therefore, we treat the relation between A and A^0 more concretely.

Our presentation of the relation between A and A^0 relies on the properties of projective representations of the time-frequency plane $\mathbb{R}^d \times \widehat{\mathbb{R}}^d$. Later we will need some results about bicharacters and 2-cocycles associated to our projective representations. Therefore, we now recall their definitions and some of their basic properties. For a similar treatment of projective representation, see [6].

Let G be a locally compact abelian group and $\mathbb{T}$ the multiplicative group of complex numbers of absolute value 1. A map β of $G \times G$ with values in $\mathbb{T}$ is a *multiplier* or (2-)*cocycle* if it satisfies for all $x, y, z \in G$:

$$\beta(x,0) = \beta(0,x) = 1,$$
$$\beta(x+y,z)\beta(x,y) = \beta(x,y+z)\beta(y,z).$$

Two cocycles β and β' are called *equivalent* or *cohomologous* if there is a Borel map c of G into $\mathbb{T}$, such that for all $x, y \in G$

$$\beta'(x,y) = \beta(x,y)\frac{c(x+y)}{c(x)c(y)}$$

A *projective representation* π of G is a map of G into the unitary group of a Hilbert space $\mathcal{H}$ such that for a cocycle β

$$\pi(x)\pi(y) = \beta(x,y)\pi(x+y), \quad \pi(0) = 1.$$

The map $X = (x,\omega) \mapsto \pi(x,\omega) = M_\omega T_x$ for $X \in \mathbb{R}^d \times \widehat{\mathbb{R}}^d$ into $\mathcal{U}(L^2(\mathbb{R}^d))$ is a projective representation of the time-frequency plane with cocycle $\beta'(X,Y) = e^{2\pi i y\cdot\omega}$ for $X = (x,\omega)$ and $Y = (y,\eta)$.

This projective representation of the time-frequency plane is equivalent to a projective representation π with the symplectic bicharacter $\beta(X,Y) = e^{2\pi i(y\omega - x\eta)}$ via the map $c : (x,\omega) \mapsto e^{2\pi i x\cdot\omega}$ for $X = (x,\omega)$ and $Y = (y,\eta)$ in $\mathbb{R}^d \times \widehat{\mathbb{R}}^d$.

Recall that a *bicharacter* of a locally compact abelian group G is a continuous map b of $G \times G$ into $\mathbb{T}$, which is a character in each argument. Any such b induces a morphism $\gamma = \gamma_b$ of G into $\widehat{G}$ by

$$\langle x, \gamma_b(y)\rangle = b(x,y).$$

A bicharacter is called *antisymmetric* if it satisfies for all $x, y \in G$

$$b(x,y)b(y,x) = 1, \quad b(x,x) = 1$$

and *symplectic* if it is alternating and γ_b is an isomorphism.

In the study of projective representations π of G with cocycle β symplectic bicharacters of G appear naturally in the commutation rule

$$\pi(x)\pi(y)\pi(x)^{-1}\pi(y)^{-1} = b(x,y)I \tag{2}$$

with $b(x,y) = \frac{\beta(x,y)}{\beta(y,x)}$ for $x, y \in G$.

For our projective representation by time-frequency shifts of $\mathbb{R}^d \times \widehat{\mathbb{R}}^d$ we recover the bicharacter $\rho(X,Y) = e^{2\pi i(y\omega - \eta x)}$ for $X = (x,\omega)$ and

$Y = (y, \eta)$. Due to the close relation of the projective representation of $\mathbb{R}^d \times \widehat{\mathbb{R}}^d$ by time-frequency shifts $\pi(x, \omega) = M_\omega T_x$ with Heisenberg's commutation relation the bicharacter $\rho(X, Y) = e^{2\pi i(y\omega - \eta x)}$ is called the *Heisenberg cocycle* for $\mathbb{R}^d \times \widehat{\mathbb{R}}^d$.

The commutation relation (2) motivates the following definition.

Definition 1: Let A be a subset of G and b a bicharacter of G. Then the *adjoint set* A_b^0 of A with respect to b is given by

$$A_b^0 = \{x \in G \ : \ b(x, a) = 1 \ \text{ for all } \ a \in A\}.$$

The adjoint of a set A is a closed subgroup of G by continuity of the character b. Furthermore a subgroup A of G is called to be *isotropic* for b if $b|_{A \times A} \equiv 1$, or equivalently $A \subset A_b^0$. We call a group A of G *maximal isotropic* for b if $A = A_b^0$, which are again closed subgroups of G.

We are now in position to answer our original question on the relation between a set A generating a non-uniform Gabor system $\mathcal{G}$ and the set A^0 generating the commutant $\mathcal{G}'$.

$$\begin{aligned} A \subset A_b^{00} = A_b^{(4)} = \cdots = \\ A_b^0 = A_b^{(3)} = A_b^{(5)} = \cdots, \end{aligned}$$

where $A_b^{(n)}$ denotes the nth adjoint of A.

The above chains of relations are the set analogous of (2) and therefore, the map $A \mapsto A_b^{(0)}$ is the desired correspondence.

For the projective representation of $\mathbb{R}^d \times \widehat{\mathbb{R}}^d$ by time-frequency shifts $\pi(x, \omega) = M_\omega T_x$ with the Heisenberg cocycle ρ we obtain the well-known *adjoint group* Λ^0 of a regular Gabor system generated by a lattice Λ of $\mathbb{R}^d \times \widehat{\mathbb{R}}^d$. In this case the maximal isotropic lattice of $\mathbb{R}^d \times \widehat{\mathbb{R}}^d$ is the standard (von Neumann) lattice $\mathbb{Z}^d \times \mathbb{Z}^d$.

Let us mention that the operator algebra $\mathcal{G} = \{\pi(\lambda) \ : \ \lambda \in \Lambda\}$ is a commutative group of time-frequency shifts if and only if Λ is isotropic, i.e. $\Lambda \subset \Lambda^0$.

It is an important fact that one can interpret the Heisenberg cocycle as follows: ρ is a character of $\mathbb{R}^d \times \widehat{\mathbb{R}}^d$, and every character of $\mathbb{R}^d \times \widehat{\mathbb{R}}^d$ is of the form

$$X \mapsto \rho(X, X') \quad \text{for some} \quad X' \in \mathbb{R}^d \times \widehat{\mathbb{R}}^d. \tag{3}$$

This induces an isomorphism between $\mathbb{R}^d \times \widehat{\mathbb{R}}^d$ and its dual group $\widehat{\mathbb{R}}^d \times \mathbb{R}^d$.

If Λ is a lattice in $\mathbb{R}^\times \widehat{\mathbb{R}}^d$, then every character of Λ extends to a character of $\mathbb{R}^d \times \widehat{\mathbb{R}}^d$ and therefore every character of Λ is of the form

$$\lambda \mapsto \rho(\lambda, Y), \qquad \lambda \in \Lambda \tag{4}$$

for some $Y \in \mathbb{R}^d \times \widehat{\mathbb{R}}^d$, where Y needs not to be unique. The homomorphism from $\mathbb{R}^d \times \widehat{\mathbb{R}}^d$ to $\widehat{\Lambda}$ has as kernel the *adjoint lattice*

$$\Lambda^0 = \{Y \in \mathbb{R}^d \times \widehat{\mathbb{R}}^d \mid \rho(\lambda, Y) = 1 \text{ for all } \lambda \in \Lambda\}. \tag{5}$$

As a consequence we have, that the adjoint set of a lattice has the structure of a lattice. The skew-bicharacter ρ of $\mathbb{R}^d \times \widehat{\mathbb{R}}^d$ gives a Fourier transform $\widehat{F}^s$ on the time-frequency plane. We call

$$\begin{aligned}\widehat{F}^s(Y) &= \iint_{\mathbb{R}^d \times \widehat{\mathbb{R}}^d} \rho(Y, X) F(X) dX \\ &= \iint_{\mathbb{R}^d \times \widehat{\mathbb{R}}^d} e^{2\pi i (y\cdot\omega - x\cdot\eta)} F(x, \omega) dx d\omega\end{aligned}$$

the *symplectic Fourier transform* of $F \in L^2(\mathbb{R}^d \times \widehat{\mathbb{R}}^d)$, since it is induced by the symplectic form Ω on $\mathbb{R}^d \times \widehat{\mathbb{R}}^d$. The symplectic Fourier transform will be essential in our proof of the Fundamental Identity of Gabor Analysis.

3. Noncommutative Tori and Feichtinger's Algebra

A *noncommutative 2d-torus* $\mathcal{A}_\Theta$ is the universal C^*-algebra generated by $2d$ unitaries $U_1, ..., U_{2d}$ subject to the commutation relations

$$U_j U_k = e^{2\pi i \theta_{jk}} U_k U_j, \qquad k, j = 1, ..., 2d$$

for a skew symmetric matrix $\Theta = \left(\theta_{jk}\right)$ with real entries.

We regard Θ as a real skew-bilinear form on $\mathbb{Z}^{2d}$, with entries given by $\Theta(e_j, e_k) = \theta_{jk}$. Then a noncommutative $2d$-torus $\mathcal{A}_\Theta$ is the twisted group C^*-algebra $C^*(\mathbb{Z}, \beta)$, where $\beta : \mathbb{Z}^{2d} \times \mathbb{Z}^{2d} \to \mathbb{T}$ is a 2-cocycle such that

$$\beta(\lambda, \lambda')\overline{\beta(\lambda', \lambda)} = e^{2\pi i \Theta(\lambda, \lambda')} \qquad \text{for} \quad \lambda, \lambda' \in \mathbb{Z}^{2d}.$$

We remark that a noncommutative 2-torus A_θ is often called *rotation algebra*. The commutation rules for the two unitary operators U and V generating A_θ read as

$$UV = e^{2\pi i \theta} VU,$$

for a real number θ. Let $\theta = \alpha\beta$ then the C^*-algebra generated by time-frequency shifts $\pi(\alpha k, \beta l) = M_{\beta l}T_{\alpha k}$ for $k, l \in \mathbb{Z}^d$ is a representation of $\mathcal{A}_\theta$ on $L^2(\mathbb{R}^d)$.

In this paper we want to treat the general case of $2d$-noncommutative tori or equivalently C^*-algebras $C^*(\Lambda, \beta)$ of time-frequency shifts $\pi(\lambda)$ for a lattice Λ in $\mathbb{R}^d \times \widehat{\mathbb{R}}^d$ with

$$\pi(\lambda)\pi(\lambda') = \beta(\lambda, \lambda')\pi(\lambda + \lambda'), \qquad \lambda, \lambda' \in \Lambda.$$

Therefore, an element of $C^*(\Lambda, \beta)$ is given by

$$\sum_{\lambda \in \Lambda} a_\lambda \pi(\lambda)$$

for a bounded complex-valued sequence $\mathbf{a} = (a_\lambda)_{\lambda \in \Lambda}$.

Moreover, this representation is **faithful** on $L^2(\mathbb{R}^d)$, which is of great importance in our proofs. One consequence, is that it is sufficient to establish statements for a dense subspace of $L^2(\mathbb{R}^d)$. For an operator algebraic proof see [35] and in [19] a proof is given using time-frequency methods.

The choice of a sequence spaces on Λ induces on the noncommutative torus an additional structure. The space $\mathcal{S}(\Lambda)$ of sequences on Λ with decay faster than the inverse of any polynomial yields a *smooth* structure on the algebra of functions on $C^*(\Lambda, \beta)$, i.e.

$$\mathcal{A}_\Lambda^\infty = \{A \in \mathcal{B}(L^2(\mathbb{R}^d)) \; : \; A = \sum_\lambda a_\lambda \pi(\lambda), \;\; \mathbf{a} = (a_\lambda)_{\lambda \in \Lambda} \in \mathcal{S}(\Lambda)\}.$$

In the present paper we introduce another structure on $C^*(\Lambda, \beta)$. Namely,

$$\mathcal{A}_\Lambda^1 = \{A \in \mathcal{B}(L^2(\mathbb{R}^d)) \; : \; A = \sum_\lambda a_\lambda \pi(\lambda), \;\; \mathbf{a} = (a_\lambda)_{\lambda \in \Lambda} \in \ell^1(\Lambda)\}. \quad (6)$$

The commutation rules for the time-frequency shifts $\pi(\lambda)$ give $C^*(\Lambda, \beta)$ the following structure:

(1) Let $A_1 = \sum_{\lambda \in \Lambda} a_1(\lambda)\pi(\lambda)$ and $A_2 = \sum_{\lambda \in \Lambda} a_2(\lambda)\pi(\lambda)$ for $\mathbf{a_1}, \mathbf{a_2} \in \ell^1(\Lambda)$ then the product of A_1 and A_2 is given by

$$A_1 \cdot A_2 = \sum_{\lambda \in \Lambda} \mathbf{a_1} \natural_\Lambda \mathbf{a_2}(\lambda)\pi(\lambda),$$

where

$$\mathbf{a_1}\natural_\Lambda\mathbf{a_2}(\lambda) = \sum_{\mu\in\Lambda} a_1(\mu)a_2(\lambda-\mu)\beta(\mu,\lambda-\mu)$$

denotes *twisted convolution* of $\mathbf{a_1}$ and $\mathbf{a_2}$ and $\mathbf{a_1}\natural_\Lambda\mathbf{a_2}(\lambda)$ is again in $\in \ell^1(\Lambda)$.

(2) Let $A = \sum_{\lambda\in\Lambda} a(\lambda)\pi(\lambda)$ for $\mathbf{a} \in \ell^1(\Lambda)$ then involution

$$A^* = \sum_{\lambda\in\Lambda} a(\lambda)^*\pi(\lambda)$$

induces an involution on $\ell^1(\Lambda)$:

$$a(\lambda)^* = \beta(\lambda,\lambda)\overline{a(-\lambda)}. \tag{7}$$

Therefore, we have that

Proposition 2: $\left(\ell^1(\Lambda), \natural_\Lambda, *\right)$ *is an involutive Banach algebra.*

For the construction of a Hilbert $C^*(\Lambda,\beta)$-module V we are looking for a time-frequency homogenous Banach space, with properties similar to $\mathcal{S}(\mathbb{R}^d)$. In his seminal paper [8] Feichtinger has introduced such a space, nowadays called *Feichtinger's algebra* and denoted by $S_0(\mathbb{R}^d)$.

Hence, we recall the definition of Feichtinger's algebra and some of its basic properties. For a detailed discussion we refer the reader to [8,12,18]. There are many characterizations of Feichtinger's algebra $S_0(\mathbb{R}^d)$. The connection between Feichtinger's algebra and Rieffel's work is most transparent in terms of time-frequency analysis.

Time-frequency representations contain information about the content of time and frequency in a signal f. For our further investigations we restrict our considerations to the *Short Time Fourier Transform* (STFT). The STFT of a function f with respect to a window function g is defined for $f, g \in L^2(\mathbb{R}^d)$ as

$$V_g f(x,\omega) = \int_{\mathbb{R}^d} f(t)\overline{g(t-x)}e^{-2\pi i t\cdot\omega}dt, \quad (x,\omega) \in \mathbb{R}^d \times \widehat{\mathbb{R}}^d \tag{8}$$

or equivalently as

$$V_g f(x,\omega) = \langle f, M_\omega T_x g\rangle, \quad \text{for } (x,\omega) \in \mathbb{R}^d \times \widehat{\mathbb{R}}^d. \tag{9}$$

Feichtinger's algebra $S_0(\mathbb{R}^d)$ is defined as follows

$$S_0(\mathbb{R}^d) = \{f \in L^2(\mathbb{R}^d) \mid \iint_{\mathbb{R}^d \times \widehat{\mathbb{R}}^d} |V_\varphi f(x,\omega)| dx d\omega < \infty\}, \tag{10}$$

where $\varphi(x) = 2^{-d/4} e^{-\pi x \cdot x}$ is a Gaussian and its norm is defined by

$$\|f\|_{S_0} = \iint_{\mathbb{R}^d \times \widehat{\mathbb{R}}^d} |V_\varphi f(x,\omega)| dx d\omega.$$

Any other non-zero Schwartz function $g \in \mathcal{S}(\mathbb{R}^d)$, instead of the Gaussian φ, yields the same space and an equivalent norm for $S_0(\mathbb{R}^d)$.

Remark 3: Feichtinger's algebra is a particular example of a class of Banach spaces, the so-called *modulation spaces*, which Feichtinger defined via integrability and decay conditions on the STFT over $\mathbb{R}^d \times \widehat{\mathbb{R}}^d$ (cf. [9]). They have been recognized as the correct class of function spaces for questions in time-frequency analysis, especially Gabor analysis, [10,12].

Theorem 4: *$S_0(\mathbb{R}^d)$ is a Banach algebra under pointwise multiplication.*

By this definition of $S_0(\mathbb{R}^d)$ elementary properties of the STFT yield invariance properties of $S_0(\mathbb{R}^d)$. In the following lemma we state two well-known facts about STFT.

Lemma 5: *Let $f, g \in L^2(\mathbb{R}^d)$ and $(u, \eta) \in \mathbb{R}^d \times \widehat{\mathbb{R}}^d$. Then*

(1) *Covariance Property of the STFT*

$$V_g(\pi(u,\eta)f)(x,\omega) = e^{2\pi i u \cdot (\omega - \eta)} V_g f(x-u, \omega-\eta).$$

(2)

$$V_g f(x,\omega) = e^{-2\pi i x \cdot \omega} V_{\hat{g}} \hat{f}(\omega, -x).$$

Proof:

(1) The covariance property of the STFT is a consequence of the commutation relation

$$T_x M_\omega = e^{-2\pi i x \cdot \omega} M_\omega T_x, \quad (x,\omega) \in \mathbb{R}^d \times \widehat{\mathbb{R}}^d,$$

furthermore one has by definition of the STFT:

$$\begin{aligned} V_g(\pi(u,\eta)f)(x,\omega) &= \langle M_\eta T_u f, M_\omega T_x g\rangle \\ &= \langle f, T_{-u} M_{-\eta} M_\omega T_x g\rangle \\ &= e^{2\pi i u\cdot(\omega-\eta)} V_g f(x-u,\omega-\eta). \end{aligned}$$

(2) The formula expresses, that for the STFT the Fourier transform yields a rotation of the time-frequency plane by an angle of 90°. It is another manifestation of the fact that the STFT contains information of f and $\hat{f}$. □

Many properties of Feichtinger's algebra $S_0(\mathbb{R}^d)$ are elementary consequences of properties of the STFT. The following theorem may be seen as a realization of this principle, where we show translation invariance and Fourier invariance of $S_0(\mathbb{R}^d)$ from Lemma 5.

Theorem 6: *Let $f \in S_0(\mathbb{R}^d)$ and $(u,\eta) \in \mathbb{R}^d \times \widehat{\mathbb{R}}^d$. Then*

(1) $\pi(u,\eta)f \in S_0(\mathbb{R}^d)$ *and* $\|f\|_{S_0} = \|\pi(u,\eta)f\|_{S_0}$.
(2) $\hat{f} \in S_0(\mathbb{R}^d)$ *and* $\|\hat{f}\|_{S_0} = \|f\|_{S_0}$.

Proof:

(1) By definition of $S_0(\mathbb{R}^d)$ we have by the invariance of the Gaussian φ under Fourier transform that

$$\begin{aligned} \|f\|_{S_0} &= \iint_{\mathbb{R}^d\times\widehat{\mathbb{R}}^d} |V_\varphi(M_\eta T_u f)(x,\omega)|dxd\omega \\ &= \iint_{\mathbb{R}^d\times\widehat{\mathbb{R}}^d} |V_\varphi f(x-u,\omega-\eta)|dxd\omega. \end{aligned}$$

(2) For the invariance under the Fourier transform we use (2) of Lemma 5, and that the definition of $S_0(\mathbb{R}^d)$ is independent of the window $g \in$

$\mathcal{S}(\mathbb{R}^d)$ and that different windows yield equivalent norms for $S_0(\mathbb{R}^d)$:

$$\begin{aligned}\|\hat{f}\|_{S_0(\mathbb{R}^d)} &= \iint_{\mathbb{R}^d\times\widehat{\mathbb{R}}^d} |V_\varphi \hat{f}(x,\omega)|dxd\omega \\ &= \iint_{\mathbb{R}^d\times\widehat{\mathbb{R}}^d} |V_{\hat{\varphi}} \hat{f}(x,\omega)|dxd\omega \\ &= \iint_{\mathbb{R}^d\times\widehat{\mathbb{R}}^d} |V_\varphi f(-\omega,x)|dxd\omega \\ &= \iint_{\mathbb{R}^d\times\widehat{\mathbb{R}}^d} |V_\varphi f(x,\omega)|dxd\omega = \|f\|_{S_0}.\end{aligned}$$ □

In our treatment of Rieffel's associativity condition for $\big(C^*(\Lambda), C^*(\Lambda^0)\big)$ we shall need the following properties of $S_0(\mathbb{R}^d)$.

Theorem 7: *Let f, g in $S_0(\mathbb{R}^d)$ then*

$$V_g f \in S_0(\mathbb{R}^d \times \widehat{\mathbb{R}}^d).$$

For a proof of this statement we refer the reader to [12], but it also follows from the functorial properties and minimality of $S_0(\mathbb{R}^d)$ in [8].

Let F be a function on the time-frequency plane $\mathbb{R}^d \times \widehat{\mathbb{R}}^d$ then the *sampling operator* for a lattice Λ in $\mathbb{R}^d \times \widehat{\mathbb{R}}^d$ is defined as follows

$$\mathbf{R}_\Lambda : F \mapsto \big(F(\lambda)\big)_{\lambda\in\Lambda}.$$

Remark 8: We will also write occasionally $F|_\Lambda$ instead of $\mathbf{R}_\Lambda F$.

Theorem 9: *Let Λ be a lattice in $\mathbb{R}^d \times \widehat{\mathbb{R}}^d$ and $F \in S_0(\mathbb{R}^d \times \widehat{\mathbb{R}}^d)$ then*

$$\mathbf{R}_\Lambda F \in S_0(\Lambda) = \ell^1(\Lambda) \tag{11}$$

and $\mathbf{R}_\Lambda$ is bounded on $S_0(\mathbb{R}^d)$.

As a consequence of the last theorem we obtain:

Corollary 10: *Let $f, g \in S_0(\mathbb{R}^d)$ then $V_g f|_\Lambda \in \ell^1(\Lambda)$, i.e.*

$$\sum_{\lambda\in\Lambda} |\langle f, \pi(\lambda)g\rangle| < \infty. \tag{12}$$

Remark 11: For Λ discrete $S_0(\Lambda) = \ell^1(\Lambda)$.

We now define a left action of $C^*(\Lambda,\beta)$ on $S_0(\mathbb{R}^d)$ by a *Gabor expansion* for the window g and the lattice $\Lambda \in \mathbb{R}^d \times \widehat{\mathbb{R}}^d$:

$$\mathbf{a}g = \sum_{\lambda\in\Lambda} a(\lambda)\pi(\lambda)g, \qquad \mathbf{a} = (a_\lambda) \in S_0(\Lambda,\beta).$$

The invariance of $S_0(\mathbb{R}^d)$ under time-frequency shifts implies that this action is well-defined on $S_0(\mathbb{R}^d)$.

Proposition 12: *Let* $\mathbf{a} \in S_0(\Lambda,\beta)$ *and* $g \in S_0(\mathbb{R}^d)$, *then*

$$\|\sum_{\lambda\in\Lambda} a(\lambda)\pi(\lambda)g\|_{S_0} \le \|\mathbf{a}\|_1 \|g\|_{S_0}.$$

Proof:

$$\begin{aligned}\|\sum_{\lambda\in\Lambda} a(\lambda)\pi(\lambda)g\|_{S_0} &\le \sum_{\lambda\in\Lambda} |a(\lambda)| \|\pi(\lambda)g\|_{S_0} \\ &= \sum_{\lambda\in\Lambda} |a(\lambda)| \|g\|_{S_0} \\ &= \|\mathbf{a}\|_1 \|g\|_{S_0}.\end{aligned}$$

□

Corollary 10 and Proposition 12 yield that for $f, g \in S_0(\mathbb{R}^d)$ the left action of

$$\langle f, g\rangle_\Lambda = V_g f|_\Lambda$$

is well-defined on $S_0(\mathbb{R}^d)$. In Gabor analysis the mapping of $\langle f, g\rangle_\Lambda \mapsto \langle f, g\rangle_\Lambda g$ is called a Gabor type frame operator with window g and lattice Λ, denoted by

$$S_{g,\Lambda} f = \sum_{\lambda\in\Lambda} \langle f, \pi(\lambda)g\rangle \pi(\lambda)g.$$

The above discussion shows that the Gabor type frame operator $S_{g,\Lambda}$ is continuous on $S_0(\mathbb{R}^d)$ for $f, g \in S_0(\mathbb{R}^d)$. In Section 5 we present some consequences for the reconstruction of square-integrable functions $f \in L^2(\mathbb{R}^d)$.

Rieffel's central observation was that

$$\langle f, g\rangle_{\mathcal{A}} = \sum_{\lambda\in\Lambda} \langle f, \pi(\lambda)g\rangle \pi(\lambda), \qquad f, g \in \mathcal{S}(\mathbb{R}^d) \tag{13}$$

is a C^*-valued innerproduct for $\mathcal{A} = C^*(\Lambda,\beta)$. In the sequel we prove that (13) defines a C^*-valued innerproduct for $C^*(\Lambda,\beta)$ for f, g in Feichtinger's algebra $S_0(\mathbb{R}^d)$.

First we show that (13) is compatible with the action of $S_0(\Lambda, \beta)$ on $S_0(\mathbb{R}^d)$. More precisely, we prove the following proposition.

Proposition 13: *Let $f, g \in S_0(\mathbb{R}^d)$ and let $\mathbf{a} \in S_0(\Lambda, \beta)$. Then*

$$\langle \mathbf{a}f, g\rangle_\Lambda = \mathbf{a}\natural_\Lambda \langle f, g\rangle_\Lambda .$$

Proof: For $\lambda \in \Lambda$ we have

$$\begin{aligned}
\langle \mathbf{a}f, \pi(\lambda)g\rangle &= \sum_{\lambda' \in \Lambda} a(\lambda')\langle \pi(\lambda')f, \pi(\lambda)g\rangle \\
&= \sum_{\lambda' \in \Lambda} a(\lambda')\langle f, \pi^*(\lambda')\pi(\lambda)g\rangle \\
&= \sum_{\lambda' \in \Lambda} a(\lambda')\langle f, \pi(\lambda - \lambda')g\rangle \beta(\lambda', \lambda - \lambda') \\
&= \mathbf{a}\natural_\Lambda \langle f, \pi(\lambda)g\rangle,
\end{aligned}$$

since $\pi(\lambda')\pi(\lambda - \lambda') = \beta(\lambda', \lambda - \lambda')\pi(\lambda)$. □

We now come to the statement of one of our main theorems.

Theorem 14: *Feichtinger's algebra $S_0(\mathbb{R}^d)$ is a left Hilbert $C^*(\Lambda, \beta)$-module with respect to the inner product, given for f, g in $S_0(\mathbb{R}^d)$ by*

$$\langle f, g\rangle_{\mathcal{A}} = \sum_{\lambda \in \Lambda} V_g f(\lambda)\pi(\lambda) .$$

The proof of Theorem 14 is postponed now and will be given after the discussion of the **F**undamental **I**dentity of **G**abor **A**nalysis, because the positivity of the innerproduct is a direct consequence of FIGA. In Section 4 we derive FIGA from an identity for products of STFT by an application of Poisson's summation formula for the symplectic Fourier transform, which requires the adjoint lattice Λ^0 of the lattice Λ in $\mathbb{R}^d \times \widehat{\mathbb{R}}^d$. In Section 4 we study the structure of $C^*(\Lambda)$ and define a $C^*(\Lambda^0)$-valued innerproduct. The proof of Rieffel's associativity condition is an elementary reformulation of the FIGA. Therefore, Section 4 is the natural place for our presentation of the FIGA.

At the end of this section we present generalizations of some notions of Hilbert spaces to Hilbert C^*-modules.

Let V be a Hilbert $\mathcal{A}$-module then a *Hilbert module map* from V to V is a linear map $T : V \to V$ that respects the module action: $T(\mathbf{a}f) = \mathbf{a}\,(T(f))$ for $\mathbf{a} \in \mathcal{A}$ and $f \in V$. The adjoint of an operator on a Hilbert space plays a central role in the study of operators on Hilbert spaces and of operator algebras, such as C^*-algebras or von Neumann algebras of operators. The following definition gives a generalization of adjoints to Hilbert $\mathcal{A}$-module.

Definition 15: Let V be a Hilbert $\mathcal{A}$-module. A map $T : V \to V$ is *adjointable* if there exists a map $T^* : V \to V$ satisfying

$$\langle Tf, g\rangle_{\mathcal{A}} = \langle f, T^* g\rangle_{\mathcal{A}}$$

for all f, g in $\mathcal{A}$. The map T^* is called the *adjoint* of T. We denote the set of all adjointable maps by $\mathcal{L}(V)$ and the set of all bounded module maps in V by $\mathcal{B}(V)$.

An elementary consequence of the definitions is the following facts about adjointable maps.

(1) Let T be in $\mathcal{L}(V)$, then its adjoint is unique and adjointable with $T^{**} = T$.
(2) Let T, S be in $\mathcal{L}(V)$, then $ST \in \mathcal{L}(V)$ with $(ST)^* = T^* S^*$.
(3) $\mathcal{L}(V)$ equipped with the operator norm $\|T\| = \sup\{\|Tx\| \;:\; \|x\| \le 1\}$ is a C^*-algebra.
(4) $\mathcal{B}(V)$ equipped with the operator norm $\|T\| = \sup\{\|Tx\| \;:\; \|x\| \le 1\}$ is a Banach algebra.

In the case of $S_0(\mathbb{R}^d)$ as $C^*(\Lambda, \beta)$-module the adjointable maps are those operators $T : S_0(\mathbb{R}^d) \to S_0(\mathbb{R}^d)$ where T^* commutes with all time-frequency shifts $\{\pi(\lambda) \;:\; \lambda \in \Lambda\}$. By definition of the $C^*(\Lambda, \beta)$-innerproduct we have

$$\begin{aligned}
\langle Tf, g\rangle_{\mathcal{A}} &= \sum_{\lambda\in\Lambda} \langle Tf, \pi(\lambda)g\rangle \pi(\lambda) \\
&= \sum_{\lambda\in\Lambda} \langle f, T^*\pi(\lambda)g\rangle \pi(\lambda) \\
&= \sum_{\lambda\in\Lambda} \langle f, \pi(\lambda)T^* g\rangle \pi(\lambda) \\
&= \langle f, T^* g\rangle_{\mathcal{A}}.
\end{aligned}$$

In [12] Feichtinger and Kozek treated selfadjoint operators on $S_0(\mathbb{R}^d)$, which commute with $\{\pi(\lambda) \;:\; \lambda \in \Lambda\}$. They called those operators Λ-

invariant. The set of all selfadjoint adjointable operators of $\mathcal{L}(S_0(\mathbb{R}^d))$ is an ideal in $\mathcal{L}(S_0(\mathbb{R}^d))$.

The notion of finite rank operators and of compact module operators is of great relevance in the construction of Morita equivalences between C^*-algebras, see Section 4.

Definition 16: Let f, g be elements of a Hilbert $\mathcal{A}$-module V. Then a **rank one operator** $K_{f,g} : V \to V$ is defined by

$$K_{f,g}h := \langle f, h \rangle_{\mathcal{A}} g.$$

The set of **compact Hilbert module operators** on V is the closed subspace of $\mathcal{L}(V)$ generated by the closure of the rank one maps $K_{f,g}$. We denote the set of compact Hilbert module operators by $\mathcal{K}(V) = \overline{\{K_{f,g} : f, g \in V\}}$.

Remark 17: A compact Hilbert module operator is not necessarily a compact operator on V, but for Hilbert $\mathbb{C}$-modules $\mathcal{H}$ the notion specializes to the definition of a compact operator on $\mathcal{H}$.

The following proposition gives some elementary facts about compact Hilbert module operators.

Proposition 18: *Let $f, g \in V$ and $T \in \mathcal{L}(V)$. Then one has*

(1) *$K_{f,g}$ is adjointable and $K^*_{f,g} = K_{g,f}$.*
(2) *$TK_{f,g} = K_{Tf,g}$.*
(3) *$K_{f,g}T = K_{f,T^*g}$.*
(4) *$\|K_{f,g}\| \leq \|f\|\|g\|$.*

A direct consequence of the previous observations is the following statement.

Proposition 19: *Let V be a Hilbert $\mathcal{A}$-module, $\mathcal{K}(V)$ is a closed ideal in $\mathcal{L}(V)$.*

Now we investigate the set of rank one module operators for our Hilbert $C^*(\Lambda, \beta)$ module $S_0(\mathbb{R}^d)$. By definition a rank one module operator is given by

$$\begin{aligned} K_{g,f}\gamma &= \langle f,\gamma\rangle_{\mathcal{A}} g \\ &= \sum_{\lambda\in\Lambda}\langle f,\pi(\lambda)\gamma\rangle\pi(\lambda)g = S_{g,\gamma,\Lambda}f, \end{aligned}$$

for $f, g, \gamma \in S_0(\mathbb{R}^d)$. The operator $S_{g,\gamma,\Lambda}$ is called a Gabor frame operator with analysis window γ and synthesis window g for a lattice Λ.

Therefore, a finite rank module operator is a finite sum of Gabor frame operators, a so-called *multi-window* Gabor frame operator. Furthermore, a rank one module operator $S_{g,\gamma,\Lambda}$ is an adjointable operator, i.e. it is a Λ-invariant operator. This elementary fact has far reaching consequences, see Section 5.

4. Feichtinger's Algebra as Bimodule for $C^*(\Lambda)$ and $C^*(\Lambda^0)$

In Section 2 we discussed the relation between an operator algebra of time-frequency shifts generated by a lattice Λ in $\mathbb{R}^d \times \widehat{\mathbb{R}}^d$. In this section we continue the discussion in the light of Morita equivalence of C^*-algebras.

The adjoint lattice of Λ in $\mathbb{R}^d \times \widehat{\mathbb{R}}^d$ was defined as the set of all points $X = (x, \omega)$ in $\mathbb{R}^d \times \widehat{\mathbb{R}}^d$ such that $\rho(\lambda, X) = 1$, which by the commutation relation of time-frequency shifts (2) is equivalent to

$$\Lambda^0 = \{\lambda^0 \in \mathbb{R}^d \times \widehat{\mathbb{R}}^d : \pi(\lambda)\pi(\lambda^0) = \pi(\lambda^0)\pi(\lambda) \quad \text{for all} \;\; \lambda \in \Lambda\}.$$

Therefore, the set of all bounded operators on $L^2(\mathbb{R}^d)$ commuting with elements from $C^*(\Lambda, \beta)$ is the C^*-algebra generated by time-frequency shifts $\pi(\lambda^0)$ for λ^0 in Λ^0.

In Section 3 we have defined a left action of $C^*(\Lambda, \beta)$ on $S_0(\mathbb{R}^d)$. Now $S_0(\mathbb{R}^d)$ has the structure of a bimodule, where the right action is induced by the opposite algebra of $C^*(\Lambda^0, \beta)$. Following Rieffel in [35], $C^*(\Lambda^0, \beta)$ can be generated by $\pi^*(\lambda^0)$ acting on the left on $S_0(\mathbb{R}^d)$, which commutes with the right action of $C^*(\Lambda, \beta)$ on $S_0(\mathbb{R}^d)$. Therefore, the opposite algebra of $C^*(\Lambda, \beta)^{\mathrm{opp}}$ is generated by $\pi^*(\lambda^0)$ with $\overline{\beta}(X, Y) = \overline{\beta(X, Y)}$ for $X = (x, \omega)$ and $Y = (y, \eta)$ as cocycle, i.e. $C^*(\Lambda^0, \overline{\beta})$. By definition the opposite algebra of $C^*(\Lambda, \beta)$ gives a right action on $S_0(\mathbb{R}^d)$ by a Gabor expansion with respect to the lattice Λ^0

$$g\mathbf{b} = |\Lambda|^{-1} \sum_{\lambda^0\in\Lambda^0} b(\lambda^0)\pi^*(\lambda^0)g, \quad g \in S_0(\mathbb{R}^d), \mathbf{b} \in S_0(\Lambda^0, \overline{\beta}).$$

Note that cohomologous cocycles yield isomorphic C^*-algebras. By a reasoning similar to the one used in Section 3 for the left action $C^*(\Lambda, \beta)$ we obtain that the right action is well-defined on $S_0(\mathbb{R}^d)$.

Before $S_0(\mathbb{R}^d)$ is given the structure of a right $C^*(\Lambda^0, \overline{\beta})$-module we state the Fundamental Identity of Gabor analysis, because it is essential in our construction of the bimodule $S_0(\mathbb{R}^d)$ for $C^*(\Lambda, \beta)$ and $C^*(\Lambda^0, \overline{\beta})$.

Theorem 20: *[FIGA] Let $f_1, g_1, f_2, g_2 \in S_0(\mathbb{R}^d)$. Then*

$$\sum_{\lambda \in \Lambda} V_{g_1} f_1(\lambda) \overline{V_{g_2} f_2(\lambda)} = |\Lambda|^{-1} \sum_{\lambda^0 \in \Lambda^0} V_{g_1} g_2(\lambda^0) \overline{V_{f_1} f_2(\lambda^0)}$$

In [35] Rieffel proved FIGA for Schwartz functions f_1, f_2, g_1, g_2 in $\mathcal{S}(G)$ for an elementary locally compact abelian group G. In [39] Tolmieri and Orr proved a special case of Rieffel's result for functions on $\mathbb{R}$ in their study of Gabor frames. Later, Janssen continued the work of Tolmieri/Orr and introduced a representation of Gabor frame operators, Janssen's representation [20]. In his proof of the Morita equivalence of $C^*(\Lambda, \beta)$ and $C^*(\Lambda^0, \overline{\beta})$ Rieffel had derived Janssen's representation of a Gabor frame operator.

Following Rieffel we use Poisson summation formula for symplectic Fourier transform in the proof of FIGA. The following theorem states the Poisson summation formula for the symplectic Fourier transform, see Section 2 for the definition.

Theorem 21: *Let $F \in S_0(\mathbb{R}^d \times \widehat{\mathbb{R}}^d)$. Then*

$$\sum_{\lambda \in \Lambda} F(\lambda) = |\Lambda|^{-1} \sum_{\lambda^0 \in \Lambda^0} \widehat{F}^s(\lambda^0) \tag{14}$$

holds pointwise and with absolute convergence of both sums.

Proof: [FIGA]

If $f, g \in S_0(\mathbb{R}^d)$ we have that $V_g f \in S_0(\mathbb{R}^d \times \widehat{\mathbb{R}}^d)$. Then $F = V_{g_1} f_1 \overline{V_{g_2} f_2}$ is in $S_0(\mathbb{R}^d \times \widehat{\mathbb{R}}^d)$, because $S_0(\mathbb{R}^d \times \widehat{\mathbb{R}}^d)$ is a Banach algebra under multiplication. Poisson's summation formula for F yields FIGA. Therefore, we compute the symplectic Fourier transform of F.

$$
\begin{aligned}
\widehat{F}^s(Y) &= \iint_{\mathbb{R}^d \times \widehat{\mathbb{R}}^d} V_{g_1} f_1(X) \overline{V_{g_2} f_2(X)} \rho(Y, X) dX \\
&= \iint_{\mathbb{R}^d \times \widehat{\mathbb{R}}^d} \langle \pi(Y) f_1, \pi(Y)\pi(X) g_1 \rangle \overline{\langle f_2, \pi(X) g_2 \rangle} \rho(X, Y) dX \\
&= \iint_{\mathbb{R}^d \times \widehat{\mathbb{R}}^d} \langle \pi(Y) f_1, \pi(X)\pi(Y) g_1 \rangle \overline{\langle f_2, \pi(X) g_2 \rangle} \rho(X, Y) dX \\
&= \langle f_1, \pi(Y) f_2 \rangle \overline{\langle g_1, \pi(Y) g_2 \rangle},
\end{aligned}
$$

where in the last step we used Moyal's formula. □

As a first application of FIGA we finish the proof of Theorem 14 by showing the positivity of

$$
\langle f, f \rangle_{\mathcal{A}} = \sum_{\lambda \in \Lambda} \langle f, \pi(\lambda) f \rangle \pi(\lambda) \tag{15}
$$

as an operator on $L^2(\mathbb{R}^d)$.

Proposition 22: *Let $f \in S_0(\mathbb{R}^d)$. Then $\langle f, f \rangle_{\mathcal{A}}$ is a positive element of $C^*(\Lambda, \beta)$.*

Proof: The representation of time-frequency shifts of $C^*(\Lambda, \beta)$ is faithful, therefore, it suffices to establish positivity for a dense subspace of $L^2(\mathbb{R}^d)$. Of course we choose $S_0(\mathbb{R}^d)$ as dense subspace. Let $g \in S_0(\mathbb{R}^d)$

$$
\begin{aligned}
\langle \langle f, f \rangle_\Lambda g, g \rangle &= \Big\langle \sum_{\lambda \in \Lambda} \langle f, \pi(\lambda) f \rangle \pi(\lambda) g, g \Big\rangle \\
&= \sum_{\lambda \in \Lambda} \langle f, \pi(\lambda) f \rangle \overline{\langle g, \pi(\lambda) g \rangle} \\
&= \sum_{\lambda^0 \in \Lambda^0} \langle f, \pi(\lambda^0) g \rangle \overline{\langle f, \pi(\lambda^0) g \rangle} \geq 0.
\end{aligned}
$$

□

In an analogous manner as in our discussion of Theorem 14 we get that the right action of $C^*(\Lambda^0, \overline{\beta})$ with properly defined $\natural_{\Lambda^0}$ and involution $*$ defines a right Hilbert $C^*(\Lambda^0, \overline{\beta})$-module structure on $S_0(\mathbb{R}^d)$ with respect to the $\mathcal{B} := C^*(\Lambda^0, \overline{\beta})$-innerproduct

$$
\langle f, g \rangle_{\mathcal{B}} := |\Lambda|^{-1} \sum_{\lambda^0 \in \Lambda^0} \langle \pi(\lambda^0) g, f \rangle \pi(\lambda^0), \qquad f, g \in S_0(\mathbb{R}^d).
$$

Two C^*-module structures $(\mathcal{A}, \langle .,. \rangle_{\mathcal{A}})$ and $(\mathcal{B}, \langle .,. \rangle_{\mathcal{B}})$ on a bimodule V are compatible if

$$\langle f, g \rangle_{\mathcal{A}} h = f \langle g, h \rangle_{\mathcal{B}}, \qquad \text{for all} \quad f, g, h \in V. \tag{16}$$

Some authors call (16) **Rieffel's associativity condition** for $\langle .,. \rangle_{\mathcal{A}}$ and $\langle .,. \rangle_{\mathcal{B}}$.

In our setting Rieffel's associativity condition expresses Janssen's representation of a Gabor frame operator $S_{g,\gamma}$ for a window $g, \gamma \in S_0(\mathbb{R}^d)$.

Theorem 23: *Let $\mathcal{A} = C^*(\Lambda, \beta)$ and $\mathcal{B} = C^*(\Lambda^0, \overline{\beta})$ with the above defined actions and innerproducts $\langle .,. \rangle_{\mathcal{A}}$ and $\langle .,. \rangle_{\mathcal{B}}$, respectively. Then*

$$S_{g,\gamma,\Lambda} f = |\Lambda|^{-1} S_{f,\gamma,\Lambda^0} g$$

for all $f, g, \gamma \in S_0(\mathbb{R}^d)$.

Proof: As in the proof of positivity of $\langle f, f \rangle_{\mathcal{A}}$ for $f \in S_0(\mathbb{R}^d)$ it suffices to show that for all $\gamma, h \in S_0(\mathbb{R}^d)$

$$\langle S_{g,\gamma,\Lambda} f, h \rangle = |\Lambda|^{-1} \langle S_{f,\gamma,\Lambda^0} g, h \rangle \,.$$

$$\begin{aligned} \langle \langle f, g \rangle_{\mathcal{A}} \gamma, h \rangle &= \sum_{\lambda \in \Lambda} \langle f, \pi(\lambda) g \rangle \overline{\langle h, \pi(\lambda) \gamma \rangle} \\ &\overset{FIGA}{=} |\Lambda|^{-1} \sum_{\lambda^0 \in \Lambda^0} \langle f, \pi(\lambda^0) h \rangle \overline{\langle g, \pi(\lambda^0) \gamma \rangle} \\ &= \big\langle f \langle g, \gamma \rangle_{\mathcal{B}}, h \big\rangle. \end{aligned}$$

□

A Hilbert C^*-module V over $\mathcal{A}$ is called *full* when the collection $\{\langle f, g \rangle_{\mathcal{A}} : \; f, g \in V\}$ is dense in $\mathcal{A}$.

Definition 24: Two C^*-algebras $\mathcal{A}$ and $\mathcal{B}$ are **strongly Morita equivalent** if there exists a full Hilbert C^*-module V over $\mathcal{B}$ such that $\mathcal{B} \simeq \mathcal{K}(V, \mathcal{A})$.

Remark 25: We denote by $\mathcal{K}(V, \mathcal{A})$ the operator closure of the finite linear combinations of "rank-one" operators $K^{\mathcal{A}}_{f,g}$.

The Morita equivalence of $C^*(\Lambda, \beta)$ and $C^*(\Lambda^0, \overline{\beta})$ is a consequence of the following theorem.

Theorem 26: *Let $S_0(\mathbb{R}^d)$ be given a bimodule structure as defined above. Let $\mathcal{A} = C^*(\Lambda, \beta)$ and $\mathcal{B} = C^*(\Lambda^0, \overline{\beta})$. Then*

(1) $\{\langle f, g\rangle_{\mathcal{A}} : \ f, g \in S_0(\mathbb{R}^d)\}$ *is dense in $\mathcal{A}$, i.e. $S_0(\mathbb{R}^d)$ is a full Hilbert $\mathcal{A}$-module.*
(2) $\{\langle f, g\rangle_{\mathcal{B}} : \ f, g \in S_0(\mathbb{R}^d)\}$ *is dense in $\mathcal{B}$, i.e. $S_0(\mathbb{R}^d)$ is a full Hilbert $\mathcal{B}$-module.*
(3) *For all $f \in S_0(\mathbb{R}^d)$ and $A \in \mathcal{A}$, we have*

$$\langle fA, fA\rangle_{\mathcal{A}} \leq \|A\|^2 \langle f, f\rangle_{\mathcal{A}},$$

i.e. boundedness of the right action.
(4) *For all $f \in S_0(\mathbb{R}^d)$ and $B \in \mathcal{B}$, we have*

$$\langle Bf, Bf\rangle_{\mathcal{B}} \leq \|B\|^2 \langle f, f\rangle_{\mathcal{B}},$$

i.e. boundedness of the left action.

implies that $S_0(\mathbb{R}^d)$ is an equivalence bimodule $(\mathcal{A}, \mathcal{B})$ with norm $\|f\| := \langle f, f\rangle_{\mathcal{A}}^{1/2}$.

Proof: Our proof follows Rieffel's approach, see [35].

(1) The linear span of the range of $\langle .,.\rangle_{\mathcal{A}}$ is an ideal in $\mathcal{A}$. Then the norm closure I of this linear span is an ideal in $\mathcal{A}$. Furthermore I is invariant under modulation and because $\pi(\lambda)$ is a faithful representation of $\mathcal{A}$, we get the desired conclusion.
(2) By similar arguments as in (1).
(3) It suffices to verify the inequality for a dense subspace of $L^2(\mathbb{R}^d)$. Let $h \in S_0(\mathbb{R}^d)$ and $A \in C^*(\Lambda, \beta)$, then

$$\begin{aligned}
\big\langle h\langle Af, Af\rangle_{\mathcal{A}}, h\big\rangle &= \big\langle \langle h, Af\rangle_{\mathcal{B}} Af, h\big\rangle \\
&= \langle Af, \langle Af, h\rangle_{\mathcal{B}} h\rangle \\
&= \big\langle Af, Af\langle h, h\rangle_{\mathcal{A}}\big\rangle \\
&= \big\langle A(f\langle h, h\rangle_{\mathcal{A}})^{1/2}, A(f\langle h, h\rangle_{\mathcal{A}})^{1/2}\big\rangle \\
&\leq \|A\|^2 \big\langle f, f\langle h, h\rangle_{\mathcal{A}}\big\rangle \\
&= \|A\|^2 \big\langle h\langle f, f\rangle_{\mathcal{A}}, h\big\rangle
\end{aligned}$$

holds for all f in $S_0(\mathbb{R}^d)$. A standard density argument yields the desired result.

(4) By similar arguments as in (3). $\square$

Corollary 27: *$C^*(\Lambda,\beta)$ and $C^*(\Lambda^0,\overline{\beta})$ are strongly Morita equivalent.*

By definition a *projective module* V is isomorphic to a direct summand of a free module $\mathcal{A}^n$ with standard basis $\{e_j\}$, i.e. there is a self-adjoint $n\times n$-matrix P with entries in $\mathcal{A}$ which is a projection, such that $V = P\mathcal{A}^n$. Rieffel proved that if $\mathcal{A}$ and $\mathcal{B}$ are unital C^*-algebras and if V is a $(\mathcal{B},\mathcal{A})$-equivalence bimodule, then V is a projective right $\mathcal{B}$-module, and a projective left $\mathcal{A}$-module. Furthermore, $\mathcal{A}$ is equivalent to the C^*-algebra $\mathcal{K}(V,\mathcal{B})$ of compact Hilbert $\mathcal{B}$-module operators.

In particular, let $\mathcal{B} = C^*(\Lambda^0,\overline{\beta})$ and let V denote the right $\mathcal{A}$-module obtained by completing $S_0(\mathbb{R}^d)$ as described earlier. Then, we have:

Theorem 28: *Feichtinger's algebra $S_0(\mathbb{R}^d)$ is a **finitely generated projective** $\mathcal{B}$-module and $\mathcal{K}(S_0(\mathbb{R}^d),\mathcal{B})$ is equivalent to $C^*(\Lambda,\beta)$.*

In [32] Rieffel made the observation that finitely generated projective C^*-modules possess a reconstruction formula in terms of a tight module frame, which is a generalization of the familiar notion of tight frames for Hilbert spaces. In a subsequent paper we discuss the connection between tight module frames for $C^*(\Lambda,\beta)$ and the characterization of $S_0(\mathbb{R}^d)$ with multi-window Gabor frames.

5. Application to Gabor Analysis: Biorthogonality Relation of Wexler-Raz

Recently, Gabor frames have been applied in various fields of mathematics, electrical engineering and signal analysis, see [13,15]. In this section we give a first glimpse of the usefulness of Rieffel's work on strong Morita equivalence of C^*-algebras generated by time-frequency shifts.

Let Λ be a lattice in $\mathbb{R}^d\times\widehat{\mathbb{R}}^d$ and $g\in L^2(\mathbb{R}^d)$ then a *Gabor system* $\mathcal{G}(g,\Lambda) := \{\pi(\lambda)g : \ \lambda\in\Lambda\}$ for a Gabor atom $g\in L^2(\mathbb{R}^d)$ is a *Gabor frame* if there are finite positive reals A, B such that

$$A\|f\|^2 \leq \sum_{\lambda\in\Lambda} |\langle f, \pi(\lambda)g\rangle|^2 \leq B\|f\|^2, \quad \text{for all} \quad f \in L^2(\mathbb{R}^d).$$

This is equivalent to invertibility and boundedness of the *Gabor frame operator*

$$S_{g,\Lambda}f = \sum_{\lambda\in\Lambda} \langle f, \pi(\lambda)g\rangle \pi(\lambda)g, \quad \text{for all} \quad f \in L^2(\mathbb{R}^d).$$

As a consequence of the invertibility of $S_{g,\Lambda}$ we have the following reconstruction formulas for $f \in L^2(\mathbb{R}^d)$

$$f = (S_{g,\Lambda})^{-1}S_{g,\Lambda}f = \sum_{\lambda\in\Lambda} \langle f, \pi(\lambda)g\rangle \pi(\lambda)(S_{g,\Lambda})^{-1}g \tag{17}$$

or

$$f = S_{g,\Lambda}(S_{g,\Lambda})^{-1}f = \sum_{\lambda\in\Lambda} \langle f, \pi(\lambda)(S_{g,\Lambda})^{-1}g\rangle \pi(\lambda)g. \tag{18}$$

The coefficients in reconstruction formulas (17) and (18) are not unique, because in general time-frequency shifts $\pi(\lambda)$ and $\pi(\lambda')$ are not linearly independent for $\lambda, \lambda' \in \Lambda$. Therefore, many researchers have investigated the set of all possible dual windows γ such that $S_{g,\gamma} = I$. Of special importance is the function $\gamma_0 := (S_{g,\Lambda})^{-1}g$, the *canonical dual window*. There are many characterizations of γ_0 in the set of all dual windows.

The Gabor frame operator $S_{g,\Lambda}$ commutes with time-frequency shifts $\{\pi(\lambda) : \lambda \in \Lambda\}$, therefore, the dual Gabor frame $\{\pi(\lambda)\gamma_0 : \lambda \in \Lambda\}$ has the structure of a Gabor frame. This observation and (17) for $(S_{g,\Lambda})^{-1}f$ yields that the inverse frame operator of a frame $\mathcal{G}(\Lambda, g)$ is given by

$$(S_{g,\Lambda})^{-1}f = S_{\gamma_0,\Lambda}f = \sum_{\lambda\in\Lambda} \langle f, \pi(\lambda)\gamma_0\rangle \pi(\lambda)\gamma_0. \tag{19}$$

Gröchenig and Leinert were motivated by a practical question on the quality of the canonical dual window $S_{g,\Lambda}^{-1}g$ of a Gabor frame $\mathcal{G}(g, \Lambda)$ generated by a Gabor atom g in $S_0(\mathbb{R}^d)$. They established that Feichtinger's algebra is a good class of Gabor atoms. Namely,

Theorem 29: [*Gröchenig-Leinert*] *Let $\mathcal{G}(g, \Lambda)$ be a Gabor frame generated by $g \in S_0(\mathbb{R}^d)$ then the canonical dual window $\gamma_0 = S_{g,\Lambda}^{-1}g$ is in $S_0(\mathbb{R}^d)$.*

We refer the reader to [19] for a proof of this deep result. In [20] Janssen had proved that for a Gabor frame $\mathcal{G}(g, \alpha\mathbb{Z} \times \beta\mathbb{Z})$ generated by a Schwartz function g the canonical dual window is also a Schwartz function under the restriction that $\alpha, \beta \in \mathbb{Q}$. Janssen had conjectured that his result is also valid for irrational lattice constants α, β. We remark that a resolution of Janssen's conjecture is a corollary of Connes result that $\mathcal{S}(\mathbb{R}^d)$ is closed under holomorphic functional calculus, [4] and [35].

In [41] Wexler/Raz characterized the set of all dual atoms with the structure of a Gabor frame for Gabor expansions on finite abelian groups. Their work had been extended to the continuous setting independently by Daubechies, H.L. Landau and Z. Landau in [7], by Janssen in [20] and by Ron and Shen [36],[37]. In the work on this problem the so-called *Janssen representation* of a Gabor frame operator was introduced in [20]. Also Feichtinger and Zimmermann considered this topic and found the minimal assumptions for the validity of Wexler-Raz's biorthogonality relation and Janssen's representation, [16]. In [12] and [16] Feichtinger and his collaborators introduced the notion of the adjoint lattice for elementary locally compact abelian groups, which Rieffel already used in his construction of equivalence bimodules between noncommutative tori, [35]. In this section we derive the result of Wexler-Raz from the Morita equivalence of $\mathbb{C}^*(\Lambda, \beta)$ and $C^*(\Lambda^0, \overline{\beta})$ and the relation between the canonical traces $\tau_\mathcal{A}$ and $\tau_\mathcal{B}$, respectively.

One of the early successes of operator algebras was the classification of all commutative C^*-algebras by Gelfand as the involutive complex-valued continuous functions over a compact space. Riesz's representation theorem for positive linear functionals of involutive complex-valued continuous functions over a compact space X yields to an extension of the Lebesgue integral. Therefore, integration of continuous functions over a compact space is considered as a trace on a commutative C^*-algebra. Therefore, traces or states on general C^*-algebras are the natural framework for non-commutative Radon measure theory.

The existence of canonical traces on $\mathcal{A} = C^*(\Lambda, \beta)$ and $\mathcal{B} = C^*(\Lambda^0, \overline{\beta})$ is one of the pleasant properties of noncommutative tori.

First we recall that a *faithful trace* $\tau_\mathcal{C}$ on a C^*-algebra $\mathcal{C}$ is a linear functional satisfying

$$\tau(I) = 1, \quad \text{for the identity operator } I \text{ of } \mathcal{C},$$
$$\tau(AB) = \tau(BA), \quad \text{for all } A, B \in \mathcal{C},$$
$$\tau(A^*A) > 0 \quad \text{for all nonzero } A \text{ in } \mathcal{C}.$$

In the case of $\mathcal{A}$ a normalized faithful trace $\tau_{\mathcal{A}}$ is given by

$$\tau_{\mathcal{A}}(\langle f, g\rangle_{\mathcal{A}}) = \langle f, g\rangle, \qquad f, g \in S_0(\mathbb{R}^d),$$

and for $\mathcal{B}$ the canonical trace $\tau_{\mathcal{B}}$ is normalized by

$$\tau_{\mathcal{B}}(\langle f, g\rangle_{\mathcal{B}}) = |\Lambda|^{-1}\langle f, g\rangle, \qquad f, g \in S_0(\mathbb{R}^d),$$

which follows from Morita equivalence of $\mathcal{A}$ and $\mathcal{B}$. This fact can be considered as a noncommutative Poisson summation formula

$$\tau_{\mathcal{A}}(\langle f, g\rangle_{\mathcal{A}}) = |\Lambda|^{-1}\tau_{\mathcal{B}}(\langle f, g\rangle_{\mathcal{B}}). \tag{20}$$

Our restriction in the following theorem to $g, \gamma \in S_0(\mathbb{R})$ is just for convenience. We refer to Gröchenig's excellent survey [18] of Gabor analysis for the general case of $g, \gamma \in L^2(\mathbb{R}^d)$.

Theorem 30: *[Wexler-Raz] Let $\mathcal{G}(g, \Lambda)$ be a Gabor system. Then the following statements are equivalent:*

(1) $S_{g,\gamma} = I$.
(2) $|\Lambda|^{-1}\langle\gamma, \pi(\lambda^0)g\rangle = \delta_{\lambda,0}$.

Proof: (2) $\Rightarrow$ (1) :

Follows from the fact that the identity of $\mathcal{A}$ is $I = \delta_{\lambda,0}$, where $\delta_{i,k}$ is the Kronecker delta. Therefore, by assumption

$$\tau_{\mathcal{B}}(S_{g,\gamma}) = \tau_{\mathcal{B}}(I) = \delta_{\lambda,0}$$

and by application of (20) we get

$$\tau_{\mathcal{B}}(I) = |\Lambda|^{-1}\tau_{\mathcal{A}}(\langle g, \gamma\rangle) = |\Lambda|^{-1}\langle\gamma, g\rangle\,.$$

The implication (1) $\Rightarrow$ (2) is trivial in the light of Rieffel's associativity condition. □

Corollary 31: *For dual functions $g, \gamma \in S_0(\mathbb{R}^d)$, the two Gabor systems $\mathcal{G}(g, \Lambda^0)$ and $\mathcal{G}(\gamma, \Lambda^0)$ are biorthogonal to each other on $L^2(\mathbb{R}^d)$.*

The proof is an elementary reformulation of Theorem 30.

Corollary 32: *A Gabor system $\mathcal{G}(g,\Lambda)$ is a tight frame if and only if $\mathcal{G}(\gamma,\Lambda^0)$ is an orthonormal system with frame bound $A = \|\Lambda\|^{-1}\|g\|^2$.*

The statement is well-known, see [18] for the elementary proof.

Corollary 33: *Let $g_1, ..., g_n, \gamma_1, ..., \gamma_n \in S_0(\mathbb{R}^d)$, then for the multi-window Gabor frame $S = \sum_{i=1}^n S_{g_i,\gamma_i}$ the following are equivalent:*

(1) $S_{g_1,\gamma_1} + \cdots + S_{g_1,\gamma_1} = I$.
(2) $|\Lambda|^{-1}(\langle\gamma_1, \pi(\lambda^0)g_1\rangle + \cdots + \langle\gamma_1, \pi(\lambda^0)g_1\rangle) = \delta_{\lambda,0}$.

The proof follows the same reasoning as for a single Gabor frame.

6. Conclusions

In the last decade operator algebra techniques have been of minor interest in Gabor analysis. But in [7],[20] and [19] deep results about Gabor frames were obtained with the help of operator algebras. We included our approach to the Wexler-Raz biorthogonality principle as an indication for the usefulness of Morita equivalence in Gabor analysis. In the following we list some topics, where our approach gives new insight, too.

(1) The original motivation for our study of Rieffel's results about Morita equivalence was the density result. There are different approaches to this important theorem [7],[12] and [1], which at the first sight seem unrelated. In [22] we show that all these approaches cover different aspects of Morita equivalence between C^*-algebras generated by time-frequency shifts with respect to a lattice in the time-frequency plane.
(2) Our interpretation of Rieffel's construction of equivalence bimodules for noncommutative tori in the notions of Gabor analysis enables us to answer the question posed by Manin on the connection between his quantum theta functions and the quantum theta vectors of Schwarz, see [23].

Other applications of Rieffel' setting yield new results on Feichtinger' conjecture and on the structure of multi-window Gabor frames, which is part of our current research.

Acknowledgment

These investigations are part of the author's Ph.D. thesis under the supervision of H.G. Feichtinger, whom I want to thank for many helpful discussions. Additionally I am greatly indebted to M.A. Rieffel for many comments on an earlier version of this paper which improved the style and presentation. The author was partially supported by the Austrian Science Foundation FWF project 14485 and by grant DOC-14482 of the Austrian Academy of Sciences.

References

1. I. Bekka, *Square integrable representations, von Neumann algebras and an application to Gabor analysis*, *J. Four. Anal. Appl.*, 10(4):325-349, 2004.
2. H. Bursztyn and A. Weinstein, *Poisson Geometry and Morita equivalence*, *arXiv:math. SG/0402347 v2*, 2004.
3. P. Casazza and M. Lammers, *Bracket products for Weyl-Heisenberg frames*, in [15], 71-98, 2003.
4. A. Connes, *$C^\star$-algebras et gèomètrie différentielle*, *C.R. Ac. Sci. Paris*, t.290:599-604, 1980.
5. I. Daubechies, *The wavelet transform, time-frequency localization and signal analysis*, *IEEE Trans. Inform.Theory*, 35:961-1005, 1990.
6. I. Daubechies, *Models for the irreducible representations of a Heisenberg group*, *Infin. Dimens. Anal. Quantum Probab. Relat. Top.*, 7(4):527- 546, 2004.
7. I. Daubechies, H. J. Landau and Z. Landau, *Gabor time-frequency lattices and the Wexler-Raz identity*, *J. Four. Anal. Appl.*, 1(4):437-478, 1995.
8. H. G. Feichtinger, *On a New Segal Algebra*, *Monatsh. Math.*, 92:269-289, 1981.
9. H. G. Feichtinger, Modulation spaces of locally compact Abelian groups. In R. Radha, editor, *Proc. Internat. Conf. on Wavelets and Applications*, pages 1–56, Chennai, January 2002, 2003.
10. H. G. Feichtinger and K. Gröchenig, *Banach spaces related to integrable group representations and their atomic decompositions. I.*, *J. Funct. Anal.*, 86(2):307-340, 1989.
11. H. G. Feichtinger and K. Gröchenig, *Banach spaces related to integrable group representations and their atomic decompositions. II.*, *Monatsh. Math.*, 108(2-3): 129-148, 1989.
12. H. G. Feichtinger and K. Gröchenig, *Gabor frames and time-frequency analysis of distributions*, *J. Funct. Anal.*, 146(2):464-495, 1997.
13. H. G. Feichtinger and W. Kozek, *Quantization of TF–lattice invariant operators on elementary LCA groups*, in [14], pp. 233–266.
14. H. G. Feichtinger and T. Strohmer, Gabor Analysis and Algorithms: Theory and Applications, Birkhäuser, Boston, 1998.

15. H. G. Feichtinger and T. Strohmer, Advances in Gabor Analysis, Birkhäuser, Boston, 2003.
16. H. G. Feichtinger and G. Zimmermann, *A Banach space of test functions for Gabor analysis*, in [14], pp. 123–170.
17. D. Gabor, *Theory of communication*, *J.IEEE(London)*, 93(III):429-457, 1946.
18. K. Gröchenig, Foundations of Time-Frequency Analysis, Birkhäuser, Boston, 2001.
19. K. Gröchenig and M. Leinert, *Wiener's lemma for twisted convolution and Gabor frames*, *J. Amer. Math.*, 1:1-17, 2004.
20. A. J. E. Janssen, *Duality and biorthogonality for Weyl-Heisenberg frames*, *J. Four. Anal. Appl.*, 1(4):403-436, 1995.
21. I. Kaplansky, *Modules over operator algebras*, *Trans. Amer. Math. Soc.*, 75:839-858, 1953.
22. F. Luef, *The density theorem in Gabor analysis: An operator algebraic approach*, preprint.
23. F. Luef, *On quantum theta functions and quantum theta vectors*, preprint.
24. K. Morita, *Duality for modules and its application to the theory of rings with minimum condition*, *Sci. Rep. Tokyo Kyoiku Daigaku Sect. A* 6:83-142, 1958.
25. J. Packer and M. A. Rieffel, *Wavelet filter functions, the matrix completion problem, and projective modules over* $C(\mathbb{T}^n)$, *J. Four. Anal. Appl.*, 9:2 , 101-116, 2003.
26. J. Packer and M. A. Rieffel, *Projective Multi-Resolution Analysis for* $L^2(\mathbb{R}^2)$, *J. Four. Anal. Appl.*, 10:5 , 439-464, 2004.
27. W. L. Paschke, *Inner product modules over* B^**-algebras*, *Trans. Amer. Math. Soc.* 182: 443-468, 1973.
28. H. Reiter, *Metaplectic groups and Segal algebras*, *Lect. Notes Math.*, 1382, Berlin, Springer Verlag, 1989.
29. M. A. Rieffel, *Morita equivalence for* C^**-algebras and* W^**-algebras*, *J. Pure Appl. Alg.*, 5: 51-96, 1974.
30. M. A. Rieffel, *Induced representations of* C^**-algebras*, *Adv. Math.*, 13:176-257, 1974.
31. M. A. Rieffel, *Strong Morita equivalence of certain transformation group* $C^\star$*-algebras*, *Math. Annalen*, 222:7-23, 1976.
32. M. A. Rieffel, C^**-algebras associated with irrational rotations*, *Pac. J. Math.*, 4:415-429, 1981.
33. M. A. Rieffel, *Von Neumann algebras associated with pairs of lattices in Lie groups*, *Math. Ann.*, 257(4):403-418, 1981.
34. M. A. Rieffel, *Morita equivalence for operator algebras*, *Proc. Symp. Pure Math.*, 93:285-298, 1982.
35. M. A. Rieffel, *Projective modules over higher-dimensional noncommutative tori*, *Can. J. Math.*, 40:257-388, 1988.
36. A. Ron and Z. Shen, *Weyl-Heisenberg frames and stable bases*, *Talk at Oberwolfach conference on Approx. Theory*, 1993.
37. A. Ron and Z. Shen, *Weyl-Heisenberg frames and Riesz bases in* $L^2(\mathbb{R}^d)$, *Duke Math. J.*, 89(2):237-282, 1997.

38. M. A. Rieffel and A. Schwarz, *Morita equivalence of multidimensional non-commutative tori*, *Int. J. Math.*, 10:289-299, 1999.
39. R. Tolmieri and R. Orr, *Poisson Summation, the ambiguity function and the theory of Weyl-Heisenberg frames*, *J. Four. Anal. Appl.*, 1(3):233-247, 1995.
40. X. Tang and A. Weinstein, *Quantization and Morita equivalence for constant Dirac structures on tori*, *e-print*, arXiv:math.QA/0305413, 2003.
41. J. Wexler and S. Raz, *Discrete Gabor Expansions*, *Signal Processing*, 21(3):207-221, 1990.
42. P. J. Wood, *Wavelets and Hilbert Modules*, *J. Four. Anal. Appl.*, 10:6, 573-598, 2004.

UNITARY MATRIX FUNCTIONS, WAVELET ALGORITHMS, AND STRUCTURAL PROPERTIES OF WAVELETS

Palle E. T. Jorgensen

Department of Mathematics, The University of Iowa
Iowa City, Iowa 52242, U.S.A.
E-mail: jorgen@math.uiowa.edu

"One cannot expect any serious understanding of what wavelet analysis means without a deep knowledge of the corresponding operator theory."

Yves Meyer*

Some connections between operator theory and wavelet analysis: Since the mid eighties, it has become clear that key tools in wavelet analysis rely crucially on operator theory. While isolated variations of wavelets, and wavelet constructions had previously been known, since Haar in 1910, it was the advent of multiresolutions, and subband filtering techniques which provided the tools for our ability to now easily create efficient algorithms, ready for a rich variety of applications to practical tasks. Part of the underpinning for this development in wavelet analysis is operator theory. This will be presented in the lectures, and we will also point to a number of developments in operator theory which in turn derive from wavelet problems, but which are of independent interest in mathematics. Some of the material will build on chapters in a new wavelet book, co-authored by the speaker and Ola Bratteli, see http://www.math.uiowa.edu/~jorgen/.

1. Introduction

While this series of four lectures will be on the subject of wavelets, the emphasis will be on some interconnections between topics in the mathe-

*[64]; see also the web page http://www.math.uiowa.edu/~jorgen/quotes.html.

matics of wavelets and other areas, both within mathematics and outside. Connections to operator theory, to quantum theory, and especially to signal processing will be studied. Concepts such as high-pass and low-pass filters have become synonymous with wavelet tools, but they have also had a significance from the very start of signal processing, for example early telephone signals over transatlantic cables. This was long before the much more recent advances in wavelets which started in the mid-1980's (as a resumption, in fact, of ideas going back to Alfred Haar [29] much earlier).

1.1. *Index of terminology in mathematics and in engineering*

Since the mid-1980's wavelet mathematics has served to some extent as a clearing house for ideas from diverse areas from mathematics, from engineering, as well as from other areas of science, such as quantum theory and optics. This makes the interdisciplinary communication difficult, as the lingo differs from field to field; even to the degree that the same term might have a different name to some wavelet practitioners from what is has to others. In recognition of this fact, Chapter 1 in the recent wavelet book [9] samples a little dictionary of relevant terms. Parts of it are reproduced here:

Terminology

- **multiresolution:** —*real world:* a set of band-pass-filtered component images, assembled into a mosaic of resolution bands, each resolution tied to a finer one and a coarser one.
  —*mathematics:* used in wavelet analysis and fractal analysis, multiresolutions are systems of closed subspaces in a Hilbert space, such as $L^2(\mathbb{R})$, with the subspaces nested, each subspace representing a resolution, and the relative complement subspaces representing the detail which is added in getting to the next finer resolution subspace.
- **matrix function:** a function from the circle, or the one-torus, taking values in a group of N-by-N complex matrices.
- **wavelet:** a function ψ, or a finite system of functions $\{\psi_i\}$, such that for some scale number N and a lattice of translation points on $\mathbb{R}$, say $\mathbb{Z}$, a basis for $L^2(\mathbb{R})$ can be built consisting of the functions $N^{\frac{j}{2}}\psi_i\left(N^j x - k\right)$, $j, k \in \mathbb{Z}$.

Then dulcet music swelled
Concordant with the life-strings of the soul;
It throbbed in sweet and languid beatings there,
Catching new life from transitory death;
Like the vague sighings of a wind at even
That wakes the wavelets of the slumbering sea...
—Shelley, *Queen Mab*

- **subband filter:** —*engineering:* signals are viewed as functions of time and frequency, the frequency function resulting from a transform of the time function; the frequency variable is broken up into bands, and up-sampling and down-sampling are combined with a filtering of the frequencies in making the connection from one band to the next.
  —*wavelets:* scaling is used in passing from one resolution V to the next; if a scale N is used from V to the next finer resolution, then scaling by $\frac{1}{N}$ takes V to a coarser resolution V_1 represented by a subspace of V, but there is a set of functions which serve as multipliers when relating V to V_1, and they are called subband filters.
- **cascades:** —*real world:* a system of successive refinements which pass from a scale to a finer one, and so on; used for example in graphics algorithms: starting with control points, a refinement matrix and masking coefficients are used in a cascade algorithm yielding a cascade of masking points and a cascade approximation to a picture.
  —*wavelets:* in one dimension the scaling is by a number and a fixed simple function, for example of the form $\underset{0\quad 1}{\sqcap}$ is chosen as the initial step for the cascades; when the masking coefficients are chosen the cascade approximation leads to a scaling function.
- **scaling function:** a function, or a distribution, φ, defined on the real line $\mathbb{R}$ which has the property that, for some integer $N > 1$, the coarser version $\varphi\left(\frac{x}{N}\right)$ is in the closure (relative to some metric) of the linear span of the set of translated functions $\dots, \varphi(x+1)$, $\varphi(x)$, $\varphi(x-1)$, $\varphi(x-2), \dots$.
- **logic gates:** —*in computation* the classical logic gates are realized as computers, for example as electronic switching circuits with two-level voltages, say high and low. Several gates have two input

voltages and one output, each one allowing switching between high and low: The output of the AND gate is high if and only if both inputs are high. The XOR gate has high output if and only if one of the inputs, but not more than one, is high.

- **qubits:** —*in physics and in computation:* qubits are the quantum analogue of the classical bits 0 and 1 which are the letters of classical computers, the qubits are formed of two-level quantum systems, electrons in a magnetic field or polarized photons, and they are represented in Dirac's formalism $|0\rangle$ and $|1\rangle$; quantum theory allows superpositions, so states $|\psi\rangle = a\,|0\rangle + b\,|0\rangle$, $a, b \in \mathbb{C}$, $|a|^2 + |b|^2 = 1$, are also admitted, and computation in the quantum realm allows a continuum of states, as opposed to just the two classical bits.
  —*mathematics:* a chosen and distinguished basis for the two-dimensional Hilbert space $\mathbb{C}^2$ consisting of orthogonal unit vectors, denoted $|0\rangle$, $|1\rangle$.
- **universality:** —*classical computing:* the property of a set of logic gates that they suffice for the implementation of every program; or of a single gate that, taken together with the NOT gate, it suffices for the implementation of every program.
  —*quantum computing:* the property of a set S of basic quantum gates that every (invertible) gate can be written as a sequence of steps using only gates from S. Usually S may be chosen to consist of one-qubit gates and a distinguished tensor gate t. An example of a choice for t is CNOT. An alternative universal one is the Toffoli gate.
  —*mathematics:* the property of a set S of basic unitary matrices that for every n and every $u \in \mathrm{U}_{2^n}(\mathbb{C})$, there is a factorization $u = s_1 s_2 \cdots s_k$, $s_i \in S$, with the understanding that the factors s_i are inserted in a chosen tensor configuration of the quantum register $\underbrace{\mathbb{C}^2 \otimes \cdots \otimes \mathbb{C}^2}_{n \text{ times}}$. Note that the factors s_i, the number k, and the configuration of the s_i's all depend on n and the gate $u \in \mathrm{U}_{2^n}(\mathbb{C})$ to be studied. The quantum wavelet algorithm (86) is an example of such a matrix u.
- **chaos:** a small variation or disturbance in the initial states or input of some system giving rise to a disproportionate, or exponentially growing, deviation in the resulting output trajectory, or output data. The term is used more generally, denoting rather drastic

forms of instability; and it is measured by the use of statistical devices, or averaging methods.

- $\mathrm{GL}_N(\mathbb{C})$**:** the *general linear group* of all complex $N \times N$ invertible matrices.
- $\mathrm{U}_N(\mathbb{C})$**:** $= \{ A \in \mathrm{GL}_N(\mathbb{C}) \mid AA^* = 1_{\mathbb{C}^N} \}$ where A^* denotes the adjoint matrix, i.e., $(A^*)_{i,j} = \bar{A}_{j,i}$.
- **transfer operator (transition operator):** —*in probability:* An operator which transforms signals s from input s_{in} to output s_{out}. The signals are represented as functions on some set E. In the simplest case, the operator is linear and given in terms of conditional probabilities $p(x,y)$. The number $p(x,y)$ may represent the probability of a transition from y to x where x and y are points in the set E. Then

$$s_{\text{out}}(x) = \sum_{y \in E} p(x,y)\, s_{\text{in}}(y).$$

—*in computation:* Let X and Y be functions on a set E, both taking values in $\{0,1\}$. Let Y be the initial state of the bit, and X the final state of the bit. If the process is governed by a probability distribution P, then the transition probabilities $p(x,y) := P(\{ X = x \mid Y = y \})$ are conditional probabilities: i.e., $p(x,y)$ is the probability of a final bit value x given an initial value y, and we have

$$P(\{X = x\}) = \sum_{y \in E} p(x,y)\, P(\{Y = y\}).$$

—*in wavelet theory:* Let $N \in \mathbb{Z}_+$, and let W be a positive function on $\mathbb{T} = \{ z \in \mathbb{C} \mid |z| = 1 \}$, for example $W = |m_0|^2$ where m_0 is some low-pass wavelet filter with N bands. (Positivity is only in the sense $W \geq 0$, nonnegative, and the function W may vanish on a subset of $\mathbb{T}$.) Then define a function p on $\mathbb{T} \times \mathbb{T}$ as follows:

$$p(z,w) = \begin{cases} \left(\frac{1}{N}\right) W(w) & \text{if } w^N = z, \\ 0 & \text{for all other values of } w. \end{cases}$$

We arrive at the transfer operator R_W, i.e., the operator transforming functions on $\mathbb{T}$ as follows:

$$s_{\text{out}}(z) = (R_W s_{\text{in}})(z) = \frac{1}{N} \sum_{w^N = z} W(w)\, s_{\text{in}}(w).$$

- **coherence:** —*in mathematics and physics:* The vectors ψ_i that make up a tight frame, one which is not an orthonormal basis, are said to be subjected to *coherence.* So coherent vector systems in Hilbert space are viewed as bases which generalize the more standard concept of orthonormal bases from harmonic analysis. A striking feature of the wavelets with compact support, which are based on scaling, is that the varieties of the two kinds of bases can be well understood geometrically. For example, the collapse of the wavelet orthogonality relations, degenerating into coherent vectors, happens on a subvariety of a lower dimension.

  More generally, coherent vectors in mathematical physics often arise with a continuous index, even if the Hilbert space is separable, i.e., has a countable orthonormal basis. This is illustrated by a vector system $\{\psi_{r,s}\}$, which should be thought of as a continuous analogue, i.e., a version where a sum gets replaced with an integral

$$C_{\psi}^{-1} \iint_{\mathbb{R}^2} \frac{dr\, ds}{r^2} \left|\langle\, \psi_{r,s} \mid f \,\rangle\right|^2 = \left\|f\right\|^2 .$$

  For more details, see also Section 3.3 of [15] and Chapter 3 of [55].

  In quantum mechanics, one talks, for example, about coherent states in connection with wavefunctions of the harmonic oscillator. Combinations of stationary wavefunctions from different energy eigenvalues vary periodically in time, and the question is which of the continuously varying wavefunctions one may use to expand an unknown function in without encountering overcompleteness of the basis. The methods of "coherent states" are methods for using these kinds of functions (which fit some problems elegantly) while avoiding the difficulties of overcompleteness. The term "coherent" applies when you succeed in avoiding those difficulties by some means or other. Of course, for students who have just learned about the classic complete orthonormal basis of stationary eigenfunctions, "coherent state" methods at first may seem like a daring relaxation of the rules of orthogonality, so that the term seems to stand for total freedom!

1.1.1. *Some background on Hilbert space*

Wavelet theory is the art of finding a special kind of basis in Hilbert space. Let $\mathcal{H}$ be a Hilbert space over $\mathbb{C}$ and denote the inner product $\langle \cdot \mid \cdot \rangle$. For us, it is assumed linear in the second variable. If $\mathcal{H} = L^2(\mathbb{R})$, then

$$\langle f \mid g \rangle := \int_{\mathbb{R}} \overline{f(x)}\, g(x)\, dx. \tag{1}$$

If $\mathcal{H} = \ell^2(\mathbb{Z})$, then

$$\langle \xi \mid \eta \rangle := \sum_{n \in \mathbb{Z}} \bar{\xi}_n \eta_n. \tag{2}$$

Let $\mathbb{T} = \mathbb{R}/2\pi\mathbb{Z}$. If $\mathcal{H} = L^2(\mathbb{T})$, then

$$\langle f \mid g \rangle := \frac{1}{2\pi} \int_{-\pi}^{\pi} \overline{f(\theta)}\, g(\theta)\, d\theta. \tag{3}$$

Functions $f \in L^2(\mathbb{T})$ have Fourier series: Setting $e_n(\theta) = e^{in\theta}$,

$$\hat{f}(n) := \langle e_n \mid f \rangle = \frac{1}{2\pi} \int_{-\pi}^{\pi} e^{-in\theta} f(\theta)\, d\theta, \tag{4}$$

and

$$\|f\|^2_{L^2(\mathbb{T})} = \sum_{n \in \mathbb{Z}} \left| \hat{f}(n) \right|^2. \tag{5}$$

Similarly if $f \in L^2(\mathbb{R})$, then

$$\hat{f}(t) := \int_{\mathbb{R}} e^{-ixt} f(x)\, dx, \tag{6}$$

and

$$\|f\|^2_{L^2(\mathbb{R})} = \frac{1}{2\pi} \int_{\mathbb{R}} \left| \hat{f}(t) \right|^2 dt. \tag{7}$$

Let J be an index set. We shall only need to consider the case when J is countable. Let $\{\psi_\alpha\}_{\alpha \in J}$ be a family of nonzero vectors in a Hilbert space $\mathcal{H}$. We say it is an *orthonormal basis* (ONB) if

$$\langle \psi_\alpha \mid \psi_\beta \rangle = \delta_{\alpha,\beta} \qquad \text{(Kronecker delta)} \tag{8}$$

and if

$$\sum_{\alpha \in J} |\langle \psi_\alpha \mid f \rangle|^2 = \|f\|^2 \qquad \text{holds for all } f \in \mathcal{H}. \tag{9}$$

If only (9) is assumed, but not (8), we say that $\{\psi_\alpha\}_{\alpha \in J}$ is a (normalized) *tight frame*. We say that it is a *frame* with *frame constants* $0 < A \leq B < \infty$ if

$$A \|f\|^2 \leq \sum_{\alpha \in J} |\langle \psi_\alpha \mid f \rangle|^2 \leq B \|f\|^2 \qquad \text{holds for all } f \in \mathcal{H}.$$

Introducing the rank-one operators $Q_\alpha := |\psi_\alpha\rangle \langle \psi_\alpha|$ of Dirac's terminology, see [9], we see that $\{\psi_\alpha\}_{\alpha \in J}$ is an ONB if and only if the Q_α's are projections and

$$\sum_{\alpha \in J} Q_\alpha = I \qquad (= \text{the identity operator in } \mathcal{H}). \tag{10}$$

It is a (normalized) tight frame if and only if (10) holds but with no further restriction on the rank-one operators Q_α. It is a frame with frame constants A and B if the operator

$$S := \sum_{\alpha \in J} Q_\alpha \tag{11}$$

satisfies

$$AI \leq S \leq BI$$

in the order of hermitian operators. (We say that operators $H_i = H_i^*$, $i = 1, 2$, satisfy $H_1 \leq H_2$ if $\langle f \mid H_1 f \rangle \leq \langle f \mid H_2 f \rangle$ holds for all $f \in \mathcal{H}$.)

Wavelets in $L^2(\mathbb{R})$ are generated by simple operations on one or more functions ψ in $L^2(\mathbb{R})$, the operations come in pairs, say scaling and translation, or phase-modulation and translations. If $N \in \{2, 3, \dots\}$ we set

$$\psi_{j,k}(x) := N^{j/2} \psi\left(N^j x - k\right) \qquad \text{for } j, k \in \mathbb{Z}. \tag{12}$$

1.1.2. *Connections to group theory*

We stress the discrete wavelet transform. But the first line in the two tables below is the continuous one. It is the only treatment we give to the continuous wavelet transform, and the corresponding *coherent vector decompositions*. But, as is stressed in [15], [55], and [54], the continuous version came first.

Summary of and variations on the resolution of the identity operator 1 in L^2 or in ℓ^2, for ψ and $\tilde{\psi}$ where $\psi_{r,s}(x) = r^{-\frac{1}{2}}\psi\left(\frac{x-s}{r}\right)$, $C_\psi = \int_{\mathbb{R}} \frac{d\omega}{|\omega|} |\hat{\psi}(\omega)|^2 < \infty$, similarly for $\tilde{\psi}$ and $C_{\psi,\tilde{\psi}} = \int_{\mathbb{R}} \frac{d\omega}{|\omega|} \overline{\hat{\psi}(\omega)}\,\hat{\tilde{\psi}}(\omega)$:

$N=2$	Overcomplete Basis	Dual Bases
continuous resolution	$C_\psi^{-1} \iint\limits_{\mathbb{R}^2} \frac{dr\,ds}{r^2} \lvert\psi_{r,s}\rangle\langle\psi_{r,s}\rvert = 1_{L^2}$	$C_{\psi,\tilde{\psi}}^{-1} \iint\limits_{\mathbb{R}^2} \frac{dr\,ds}{r^2} \lvert\psi_{r,s}\rangle\langle\tilde{\psi}_{r,s}\rvert = 1_{L^2}$
discrete resolution	$\sum_{j\in\mathbb{Z}} \sum_{k\in\mathbb{Z}} \lvert\psi_{j,k}\rangle\langle\psi_{j,k}\rvert = 1_{L^2}$, $\psi_{j,k}$ corresponding to $r = 2^{-j}$, $s = k2^{-j}$	$\sum_{j\in\mathbb{Z}} \sum_{k\in\mathbb{Z}} \lvert\psi_{j,k}\rangle\langle\tilde{\psi}_{j,k}\rvert = 1_{L^2}$
$N \geq 2$	Isometries in ℓ^2	Dual Operator System in ℓ^2
sequence spaces	$\sum_{i=0}^{N-1} S_i S_i^* = 1_{\ell^2}$, where $S_0, \dots, S_{N-1}$ are adjoints to the quadrature mirror filter operators F_i, i.e., $S_i = F_i^*$	$\sum_{i=0}^{N-1} S_i \tilde{S}_i^* = 1_{\ell^2}$, for a dual operator system $S_0, \dots, S_{N-1}$, $\tilde{S}_0, \dots, \tilde{S}_{N-1}$

Consult Chapter 3 of [55] for the continuous resolution, and Section 2.2 of [9] for the discrete resolution. If h, k are vectors in a Hilbert space $\mathcal{H}$, then the operator $A = |h\rangle\langle k|$ is defined by the identity $\langle u \mid Av \rangle = \langle u \mid h \rangle \langle k \mid v \rangle$ for all $u, v \in \mathcal{H}$. Then the assertions in the first table amount to:

$C_\psi^{-1} \iint\limits_{\mathbb{R}^2} \frac{dr\,ds}{r^2} \lvert\langle \psi_{r,s} \mid f \rangle\rvert^2 = \lVert f \rVert_{L^2}^2 \quad \forall f \in L^2(\mathbb{R})$	$C_{\psi,\tilde{\psi}}^{-1} \iint\limits_{\mathbb{R}^2} \frac{dr\,ds}{r^2} \langle f \mid \psi_{r,s} \rangle \langle \tilde{\psi}_{r,s} \mid g \rangle = \langle f \mid g \rangle \quad \forall f, g \in L^2(\mathbb{R})$
$\sum_{j\in\mathbb{Z}} \sum_{k\in\mathbb{Z}} \lvert\langle \psi_{j,k} \mid f \rangle\rvert^2 = \lVert f \rVert_{L^2}^2 \quad \forall f \in L^2(\mathbb{R})$	$\sum_{j\in\mathbb{Z}} \sum_{k\in\mathbb{Z}} \langle f \mid \psi_{j,k} \rangle \langle \tilde{\psi}_{j,k} \mid g \rangle = \langle f \mid g \rangle \quad \forall f, g \in L^2(\mathbb{R})$
$\sum_{i=0}^{N-1} \lVert S_i^* c \rVert^2 = \lVert c \rVert^2 \quad \forall c \in \ell^2$	$\sum_{i=0}^{N-1} \langle S_i^* c \mid \tilde{S}_i^* d \rangle = \langle c \mid d \rangle \quad \forall c, d \in \ell^2$

A function ψ satisfying the resolution identity is called a *coherent vector* in mathematical physics. The representation theory for the $(ax+b)$-group, i.e., the matrix group $G = \{ \left(\begin{smallmatrix} a & b \\ 0 & 1 \end{smallmatrix}\right) \mid a \in \mathbb{R}_+,\ b \in \mathbb{R} \}$, serves as its underpinning. Then the tables above illustrate how the $\{\psi_{j,k}\}$ wavelet system arises from a discretization of the following unitary representation of G:

$$\left(U_{\left(\begin{smallmatrix} a & b \\ 0 & 1 \end{smallmatrix}\right)} f\right)(x) = a^{-\frac{1}{2}} f\left(\frac{x-b}{a}\right) \tag{13}$$

acting on $L^2(\mathbb{R})$. This unitary representation also explains the discretization step in passing from the first line to the second in the tables above. The functions $\{ \psi_{j,k} \mid j,k \in \mathbb{Z} \}$ which make up a wavelet system result from the choice of a suitable coherent vector $\psi \in L^2(\mathbb{R})$, and then setting

$$\psi_{j,k}(x) = \left(U_{\left(\begin{smallmatrix} 2^{-j} & k \cdot 2^{-j} \\ 0 & 1 \end{smallmatrix}\right)} \psi\right)(x) = 2^{\frac{j}{2}} \psi\left(2^j x - k\right). \tag{14}$$

Even though this representation lies at the historical origin of the subject of wavelets (see [16]), the $(ax+b)$-group seems to be now largely forgotten in the next generation of the wavelet community. But Chapters 1–3 of [15] still serve as a beautiful presentation of this (now much ignored) side of the subject. It also serves as a link to mathematical physics and to classical analysis.

Since the representation U in (13) on $L^2(\mathbb{R})$, when a unitary U is defined from (13) setting $a = 2$, $b = 0$, $(Uf)(x) := 2^{-\frac{1}{2}} f\left(\frac{x}{2}\right)$, leaves invariant the Hardy space

$$\mathcal{H}_+ = \left\{ f \in L^2(\mathbb{R}) \mid \operatorname{supp}(\hat{f}) \subset [0,\infty\rangle \right\}, \tag{15}$$

formula (14) suggests that it would be simpler to look for wavelets in $\mathcal{H}_+$. After all, it is a smaller space, and it is natural to try to use the causality features of $\mathcal{H}_+$ implied by the support condition in (15). Moreover, in the world of the Fourier transform, the two operations of the formulas (13) and (14) take the simpler forms

$$\hat{f} \longmapsto a^{\frac{1}{2}} e^{-ibt} \hat{f}(at) \quad \text{and} \quad \hat{\psi} \longmapsto 2^{\frac{j}{2}} e^{-i2^j kt} \hat{\psi}\left(2^j t\right). \tag{16}$$

So in the early nineties, this was an open problem in the theory, i.e., whether or not there are wavelets in the Hardy space; but it received a beautiful answer in [1]. Auscher showed that there are no wavelet functions ψ in $\mathcal{H}_+$

which satisfy the following mild regularity properties:

(R_0) $\hat{\psi}$ is continuous;

(R_ε) for some $\varepsilon \in \mathbb{R}_+$, $\hat{\psi}(t) = \mathcal{O}(|t|^\varepsilon)$
and $\hat{\psi}(t) = \mathcal{O}\left((1+|t|)^{-\varepsilon-\frac{1}{2}}\right)$, $t \in \mathbb{R}$.

Comparison of formulas (13) and (14) shows that the traditional discrete wavelet transform may be viewed as the restriction to a subgroup H of a classical unitary representation of G. The unitary representations of G are completely understood: the set of irreducible unitary representations consists of two infinite-dimensional inequivalent subrepresentations of the representation (13) on $L^2(\mathbb{R})$, together with the one-dimensional representations $\left(\begin{smallmatrix} a & b \\ 0 & 1 \end{smallmatrix}\right) \to a^{ik}$ parameterized by $k \in \mathbb{R}$. (The two subrepresentations of (13) are obtained by restricting to $f \in L^2(\mathbb{R})$ with $\operatorname{supp} \hat{f} \subseteq (-\infty, 0]$ and $\operatorname{supp} \hat{f} \subseteq [0, \infty\rangle$, respectively.) However, the subgroup H of G has a rich variety of inequivalent infinite-dimensional representations that do not arise as restrictions of (13), or of any representation of G. The group H considered in (14) is a semidirect product (as is G): it is of the form

$$H_N = \left\{ \begin{pmatrix} a & b \\ 0 & 1 \end{pmatrix} \;\middle|\; a = N^j,\ b = \sum_{i \in \mathbb{Z}} n_i N^i,\ j \in \mathbb{Z},\ n_i \in \mathbb{Z}, \text{ where the } \sum_i \text{ summation is finite} \right\}. \tag{17}$$

(In the jargon of pure algebra, the nonabelian group H_N is the semidirect product of the two abelian groups $\mathbb{Z}$ and $\mathbb{Z}\left[\frac{1}{N}\right]$, with a naturally defined action of $\mathbb{Z}$ on $\mathbb{Z}\left[\frac{1}{N}\right]$.)

The papers [14], [45], [5], [30], [60], and [10] show that it is possible to use these nonclassical representations of H for the construction of unexpected classes of wavelets, the wavelet sets being the most notable ones. Recall that a subset $E \subset \mathbb{R}$ of finite measure is a *wavelet set* if $\hat{\psi} = \chi_E$ is such that, for some $N \in \mathbb{Z}_+$, $N \geq 2$, the functions $\left\{ N^{\frac{j}{2}} \psi(N^j x - k) \mid j, k \in \mathbb{Z} \right\}$ form an orthonormal basis for $L^2(\mathbb{R})$. Until the work of Larson and others, see [14] and [30], it was not even clear that wavelet sets E could exist in the case $N > 2$. The paper [60] develops and extends the representation theory for the subgroups H_N independently of the ambient group G and shows that each H_N has continuous series of representations which account for the wavelet sets. The role of the representations of the groups H_N and their generalizations for the study of wavelets was first stressed in [10].

There is a different transform which is analogous to the wavelet transform of (13)–(14), but yet different in a number of respects. It is the Gabor transform, and it has a history of its own. Both are special cases of the following construction: Let G be a nonabelian matrix group with center C, and let U be a unitary irreducible representation of G on the Hilbert space $L^2(\mathbb{R})$. When $\psi \in L^2(\mathbb{R})$ is given, we may define a transform

$$(T_\psi f)(\xi) := \langle U(\xi)\psi \mid f \rangle, \quad \text{for } f \in L^2(\mathbb{R}) \text{ and } \xi \in G/C. \tag{18}$$

It turns out that there are classes of matrix groups, such as the $ax+b$ group, or the 3-dimensional group of upper triangular matrices, which have transforms T_ψ admitting effective discretizations. This means that it is possible to find a vector $\psi \in L^2(\mathbb{R})$, and a discrete subgroup $\Lambda \subset G/C$, such that the restriction to Λ of the transform T_ψ in (18) is injective from $L^2(\mathbb{R})$ into functions on Λ.

There are many books on transform theory, and here we are only making the connection to wavelet theory. The book [66] contains much more detail on the group-theoretic approach to these continuous and discrete coherent vector transforms.

1.1.3. *Some background on matrix functions in mathematics and in engineering*

One of our coordinates for the landscape of multiresolution wavelets takes the form of a geometric index. In fact, it involves a traditional operator-theoretic index with values in $\mathbb{Z}$. When it is identified with a winding number or a counting of homotopy classes, it serves also as a Fredholm index of an associated Toeplitz operator. An orthogonal dyadic wavelet basis has its wavelet function ψ satisfying the normalization $\|\psi\|_{L^2(\mathbb{R})} = 1$, i.e., ψ is a vector of norm one in the Hilbert space $L^2(\mathbb{R})$. In the lingo of quantum theory, ψ is therefore a pure state, and the x-coordinate is an observable called the position. The integral $E_\psi(x) = \int_{\mathbb{R}} x\,|\psi(x)|^2\,dx$ is the expected value of the position. If ψ_H denotes the standard Haar function in (54), then clearly $E_{\psi_H}(x) = \frac{1}{2}$. Also note the translation formula $E_{\psi(\,\cdot\,-k)}(x) = E_\psi(x) + k$. We showed in Corollary 2.4.11 of [9], completely generally, that the other orthonormal wavelets ψ have expected values in the set $\frac{1}{2}+\mathbb{Z}$. Hence, after ψ is translated by an integer, you cannot distinguish it from the Haar wavelet ψ_H in (54) by looking only at the expected value of its position coordinate. The translation integer k turns out to be a winding number. Our result holds more generally when the definition of $E_\psi(x)$ is adapted to a wider wavelet context, as we showed in Chapter 6 of [9]; but in all cases, there is a winding number which produces the above-mentioned integer translate k.

The issue of connectedness for various classes of wavelets is a general question which has been addressed previously in the wavelet literature; see, e.g., [30], [35], [83], and [68]. Here we bring homotopy to bear on the question, and we identify the connected components when the compact support is fixed and given. We show among other things that for a fixed K_1-class a homotopy may take place within a variety of wavelets which is specified by a slightly bigger support than the initially given one.

An important point of our present discussion, beyond the mere fact of compact support, is the size of the support of the wavelets in question. Consider two wavelets A and B of a certain support size. Then our first results in this section also specify the paths $C(t)$, if any, which connect A and B, and in particular the size of the support of the wavelets corresponding to $C(t)$. In [9], we treat connectivity in the wider context of noncompactly supported wavelets, following at the outset [28], which considers scale number $N = 2$, and wavelets ψ satisfying

$$\left\{ 2^{\frac{j}{2}} \psi\left(2^j x - k\right) \right\}_{j,k\in\mathbb{Z}} \text{ is an orthonormal basis (ONB) for } L^2(\mathbb{R}). \tag{19}$$

Garrigós considers, for $\frac{1}{2} < \alpha \leq \infty$, the class $\mathcal{W}_\alpha$ of wavelets ψ such that

$$\int_{\mathbb{R}} |\psi(x)|^2 \left(1 + |x|^2\right)^\alpha dx < \infty, \tag{20}$$

and there is an $\varepsilon = \varepsilon(\psi)$ such that

$$\int_{\mathbb{R}} \left|\hat{\psi}(t)\right|^2 \left(1 + |t|^2\right)^\varepsilon dt < \infty, \tag{21}$$

i.e., the wavelet is supposed to have some degree of smoothness in the sense of Sobolev.

We now turn to the group of functions $U\colon \mathbb{T} \to \mathrm{U}(N)$, where $\mathrm{U}(N)$ denotes the group of all complex N-by-N matrices. The functions will not be assumed continuous in general. The continuous functions will be designated $C(\mathbb{T}, \mathrm{U}(N))$. Each function in $C(\mathbb{T}, \mathrm{U}(N))$ has a K_1-class, also called a winding number; see [9]. The functions in $C(\mathbb{T}, \mathrm{U}(N))$ with finite Fourier expansion will be called *Fourier polynomials*, also if they are functions which take values in $\mathrm{U}(N)$.

Proposition 1: *Let $U \in C(\mathbb{T}, \mathrm{U}(N))$ be a Fourier polynomial, and assume that $K_1(U) = d \in \mathbb{Z}$. Then U is homotopic in $C(\mathbb{T}, \mathrm{U}(N))$ to*

$$V(z) = z^d p \oplus (1_N - p) \tag{22}$$

where p is the one-dimensional projection onto the first coordinate slot in $\mathbb{C}^N$, and if U has the form

$$U(z) = \sum_{k=-D}^{D} z^k a_k, \tag{23}$$

then U may be homotopically deformed to V in $C(\mathbb{T}, \mathrm{U}(N))$ through Fourier polynomials of degree at most $|d| + ND$.

This proposition remains true if the word "Fourier polynomial" is replaced by "polynomial" and $a_k = 0$ for $k = -D, -D+1, \dots, -1$. In that case $d \in \mathbb{Z}_+$ and U may be homotopically deformed to V in the loop semigroup of polynomial unitaries in $C(\mathbb{T}, \mathrm{U}(N))$ through polynomials of degree at most d.

Proof: Multiplying U by z^D, we obtain a polynomial $z^D U(z)$ of degree $2D$ mapping $\mathbb{T}$ into $\mathrm{U}(N)$. Then $K_1(z^D U) = d + ND$. By Proposition 3.3 of [8], there exist $d + ND$ one-dimensional projections $p_1, p_2, \dots, p_{d+ND}$ in $M_N(\mathbb{C})$ and a unitary $V_0 \in M_N(\mathbb{C})$ such that

$$z^D U(z) = V_0 \prod_{k=1}^{d+ND} (1 - p_i + z p_i). \tag{24}$$

(See § 2.2.4 for a related, but different, decomposition.) Now, deforming each of the p_i's continuously through one-dimensional projections to the projection p_0 onto the first coordinate direction, and deforming V_0 in $\mathrm{U}(N)$ into 1_N, we see that $z^D U(z)$ can be deformed into

$$\prod_{k=1}^{d+ND} (1 - p_0 + z p_0) = 1 - p_0 + z^{d+ND} p_0. \tag{25}$$

Thus $U(z)$ itself is deformed into

$$z^{-D}(1 - p_0) + z^{d+(N-1)D} p_0. \tag{26}$$

But writing $(1 - p_0)$ as a sum of $N - 1$ one-dimensional projections $q_1, \dots, q_{N-1}$, we have that the unitary that $U(z)$ is deformed into is

$$\prod_{k=1}^{N-1} \left((1 - q_k) + z^{-D} q_k\right) \cdot \left(1 + z^{d+(N-1)D} p_0\right), \tag{27}$$

and next deforming each of the q_k in this decomposition into p_0, we see that $U(z)$ is deformed into

$$\prod_{k=1}^{N-1} \left((1 - p_0) + z^{-D} p_0 \right) \cdot \left(1 + z^{d+(N-1)D} p_0 \right) = (1 - p_0) + z^d p_0. \quad (28)$$

The crude estimate $|d| + ND$ on the degree of the Fourier polynomials occurring during the deformation is straightforward.

To prove the last statement in the proposition one does not need to multiply U by z^D, and the proof simplifies. Note in particular that $D \leq d$ (assuming $a_D \neq 0$). □

Remark 2: We do not know if Proposition 1 is true if $C(\mathbb{T}, \mathrm{U}(N))$ is replaced by $C(\mathbb{T}, \mathrm{GL}(N))$. It is known from Lemma 11.2.12 of [77] that if $A \in C(\mathbb{T}, \mathrm{GL}(N))$ is a polynomial of degree 1 in z, then A can be homotopically deformed through first-order polynomials in $C(\mathbb{T}, \mathrm{GL}(N))$ to a unitary of the form $z \to zp + (1_N - p)$ for some projection p, and hence Proposition 1 for $C(\mathbb{T}, \mathrm{GL}(N))$ would follow if any polynomial $A \in C(\mathbb{T}, \mathrm{GL}(N))$ could be factored into first-order polynomials. It is also clear, since any element $A \in C(\mathbb{T}, \mathrm{GL}(N))$ can be homotopically deformed into $z^d p \oplus (1_N - p)$ in $C(\mathbb{T}, \mathrm{GL}(N))$, that if A is a Fourier polynomial, then A can be homotopically deformed into $z^d p \oplus (1_N - p)$ through Fourier polynomials. This follows by compactness and the Stone–Weierstraß theorem (Lemma 11.2.3 of [77]). For our purposes in wavelet theory, though, we would need a computable upper bound for the degree of the Fourier polynomials.

For ease of reference we will now list the correspondences between the various objects that interest us in this case. These objects are:

(i) matrix functions, $A\colon \mathbb{T} \to \mathrm{U}_N(\mathbb{C})$, satisfying the normalization

$$A(1) = H, \qquad H_{k,l} = \frac{1}{\sqrt{N}} e^{i2\pi kl/N}, \; k,l = 0, \dots, N-1, \quad (29)$$

(ii) high- and low-pass wavelet filters m_i, $i = 0, 1, \dots, N-1$, satisfying

$$\sum_{w^N = z} \overline{m_i(w)}\, m_j(w) = N\delta_{ij}, \qquad i,j = 0, \dots, N-1, \quad (30)$$

and

$$m_0(1) = \sqrt{N}, \quad (31)$$

(iii) scaling functions φ together with wavelet generators ψ_i.

We did not specify the continuity and regularity requirements of the functions A, m_i, φ, ψ_i above. This will be done differently in different contexts and the classes clearly depend on these added requirements. We will now restrict to the case that the functions φ and ψ_i have compact support in $[0, \infty\rangle$, i.e., that A and m_i are polynomials in z. Thus $z \to A(z)$ is a polynomial function with

$$(A(z))^* A(z) = 1, \qquad z \in \mathbb{T}. \tag{32}$$

Scaling functions/wavelet generators to wavelet filters $(\varphi, \psi) \mapsto m$

One defines a_n by

$$\varphi(x) = \sqrt{N} \sum_{n \in \mathbb{Z}} a_n \varphi(Nx - n), \tag{33}$$

(cf. (107)) and then m_0 by

$$m_0(z) = \sum_n a_n z^n, \tag{34}$$

or one uses

$$\sqrt{N} \hat{\varphi}(Nt) = m_0(t) \hat{\varphi}(t) \tag{35}$$

directly. Then the high-pass filters m_i, $i = 1, \dots, N-1$, can be derived from (110) below. If we are in the generic case (106), we may also recover the Fourier coefficients $a_n^{(i)}$ of m_i by

$$\begin{aligned} a_n^{(i)} &= \left(1/\sqrt{N}\right) \langle \varphi(\cdot - n) \mid \psi_i(\cdot / N) \rangle \\ &= \langle \varphi(\cdot - n) \mid U\psi_i \rangle \qquad (\text{with } \psi_0 = \varphi), \end{aligned}$$

where $U\psi_i(x) := N^{-1/2} \psi_i(x/N)$. In particular it follows in this generic case that if the scaling and wavelet functions have compact support and the filters are Lipschitz, then the filters are Fourier polynomials. Is this true also in the nongeneric tight frame case?

Now, if $D \in \mathbb{N}$, define:

- $\mathrm{MF}(D) =$ the set of polynomial functions in $z \in \mathbb{T}$ in $C(\mathbb{T}, \mathrm{U}_N(\mathbb{C}))$ of degree at most D satisfying (29); (36)
- $\mathrm{WF}(D) =$ the set of N-tuples of wavelet filters $(m_0, \dots, m_{N-1})$ such that all m_i are polynomials in $z \in \mathbb{T}$ of degree at most D satisfying (30) and (31); (37)
- $\mathrm{SF}(D) =$ the set of N-tuples $(\varphi, \psi_1, \dots, \psi_{N-1})$ of scaling functions/ wavelet functions with support in $[0, D]$. (38)

The spaces $\mathrm{MF}(D)$, $\mathrm{WF}(D)$, and $\mathrm{SF}(D)$ may be equipped with the obvious topologies, coming in the first two cases from, for example, the L^{∞}-norm over z, and in the last case either from the $L^2(\mathbb{R})$-norm or, as will be more relevant, the tempered-distribution topology. By virtue of Proposition 3.2 in [8], $\mathrm{MF}(D)$ has the structure of a compact algebraic variety, and so by (104) below, $\mathrm{WF}(D)$ is a compact algebraic variety. It is clear from (104) that the map $A \to m$ maps $\mathrm{MF}(D)$ into $\mathrm{WF}((D+1)N-1)$, and that $m \to A$ maps $\mathrm{WF}((D+1)N-1)$ into $\mathrm{MF}(D)$. Furthermore, it is clear from (33) and (110) that $m \to (\varphi, \psi)$ maps $\mathrm{WF}((N-1)D)$ into $\mathrm{SF}(D)$, and conversely $(\varphi, \psi) \to m$ maps $\mathrm{SF}(D)$ into $\mathrm{WF}((N-1)D)$.

Now, let a subindex 0 denote the subsets of these various spaces such that the condition

$$\operatorname{Spec}(R_0) \cap \mathbb{T} = \{1\} \quad \text{and} \quad \dim\left\{g \in \mathcal{K}_{\lfloor \frac{D}{N-1} \rfloor},\ R(g) = g\right\} = 1 \qquad (39)$$

holds. It is known that the set of points such that (39) does not hold is a lower-dimensional subvariety of the various varieties, see Section 6 of [46], and hence $\mathrm{MF}_0(D)$, $\mathrm{WF}_0(D)$, and $\mathrm{SF}_0(D)$ contain the generic points in $\mathrm{MF}(D)$, $\mathrm{WF}(D)$, and $\mathrm{SF}(D)$.

We now summarize the local connectivity results by stating the following theorem. The proof may be found in [9], where this is Theorem 2.1.3.

Theorem 3: *Let $k \in \mathbb{N}$. Equip the space $\mathrm{SF}(kN+1)$ of scaling functions/wavelet functions with support in $[0, kN+1]$ with the tempered-distribution topology. Then $\mathrm{SF}(kN+1)$ is homeomorphic to a compact algebraic variety. Furthermore, for two elements $(\varphi_0, \psi_0), (\varphi_1, \psi_1) \in \mathrm{SF}(kN+1)$, the following conditions are equivalent:*

(*a*) *The elements (φ_0, ψ_0) and (φ_1, ψ_1) can be connected to each other by a continuous path in $\mathrm{SF}(NkN+1)$;*
(*b*) $K_1(\varphi_0, \psi_0) = K_1(\varphi_1, \psi_1)$;
(*c*) *The elements (φ_0, ψ_0) and (φ_1, ψ_1) can be connected to each other by a continuous path in some $\mathrm{SF}(K)$.*

Thus, $\mathrm{SF}(kN+1)$ is divided into $Nk(N-1)+1$ components which are connected over $\mathrm{SF}(NkN+1)$.

1.2. *Motivation*

In addition to the general background material in the present section, the reader may find a more detailed treatment of some of the current research trends in wavelet analysis in the following papers: [47] (a book review), [48] (a survey), and the research papers [19], [20], [21], [22], [49], and [50].

As a mathematical subject, the theory of wavelets draws on tools from mathematics itself, such as harmonic analysis and numerical analysis. But in addition there are exciting links to areas outside mathematics. The connections to electrical and computer engineering, and to image compression and signal processing in particular, are especially fascinating. These interconnections of research disciplines may be illustrated with the two subjects (1) wavelets and (2) subband filtering [from signal processing]. While they are quite different, and have distinct and independent lives, and even have different aims, and different histories, they have in recent years found common ground. It is a truly amazing success story. Advances in one area have helped the other: subband filters are absolutely essential in wavelet algorithms, and in numerical recipes used in subdivision schemes, for example, and especially in JPEG 2000—an important and extraordinarily successful image-compression code. JPEG uses nonlinear approximations and harmonic analysis in spaces of signals of bounded variation. Similarly, new wavelet approximation techniques have given rise to the kind of data-compression which is now used by the FBI [via a patent held by two mathematicians] in digitizing fingerprints in the U.S. It is the happy marriage of the two disciplines, signal processing and wavelets, that enriches the union of the subjects, and the applications, to an extraordinary degree. While the use of high-pass and low-pass filters has a long history in signal processing, dating back more than fifty years, it is only relatively recently, say the mid-1980's, that the connections to wavelets have been made. Multiresolutions from optics are the bread and butter of wavelet algorithms, and they in turn thrive on methods from signal processing, in the quadrature mirror filter construction, for example. The effectiveness of multiresolutions in data compression is related to the fact that multiresolutions are modeled on the familiar positional number system: the digital, or dyadic, representation of numbers. Wavelets are created from scales of closed subspaces of the Hilbert space $L^2(\mathbb{R})$ with a scale of subspaces corresponding to the progression of bits in a number representation. While oversimplified here, this is the key to the use of wavelet algorithms in digital representation of signals and images. The digits in the classical number representation in fact are quite analogous to the frequency subbands that are used *both* in signal processing and in wavelets.

The two functions

$$\varphi(x) = \begin{cases} 1 & 0 \le x < 1 \\ 0 & \text{elsewhere} \end{cases} \quad \text{and } \psi(x) = \begin{cases} 1 & 0 \le x < \frac{1}{2} \\ -1 & \frac{1}{2} \le x < 1 \\ 0 & \text{elsewhere} \end{cases}$$

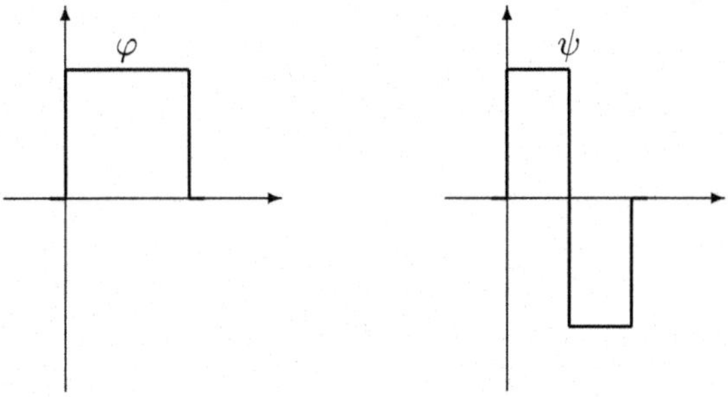

φ ψ (40)

Father function
(a)

Mother function
(b)

capture in a glance the refinement identities

$$\varphi(x) = \varphi(2x) + \varphi(2x-1) \qquad \text{and} \qquad \psi(x) = \varphi(2x) - \varphi(2x-1).$$

The two functions are clearly orthogonal in the inner product of $L^2(\mathbb{R})$, and the two closed subspaces $\mathcal{V}_0$ and $\mathcal{W}_0$ generated by the respective integral translates

$$\{\varphi(\cdot - k) : k \in \mathbb{Z}\} \qquad \text{and} \qquad \{\psi(\cdot - k) : k \in \mathbb{Z}\} \tag{41}$$

satisfy

$$U\mathcal{V}_0 \subset \mathcal{V}_0 \qquad \text{and} \qquad U\mathcal{W}_0 \subset \mathcal{V}_0 \tag{42}$$

where U is the dyadic scaling operator $Uf(x) = 2^{-1/2} f(x/2)$. The factor $2^{-1/2}$ is put in to make U a unitary operator in the Hilbert space $L^2(\mathbb{R})$. This version of Haar's system naturally invites the question of what other pairs of functions φ and ψ with corresponding orthogonal subspaces $\mathcal{V}_0$ and $\mathcal{W}_0$ there are such that the same invariance conditions (42) hold. The invariance conditions hold if there are coefficients a_k and b_k such that the scaling identity

$$\varphi(x) = \sum_{k \in \mathbb{Z}} a_k \varphi(2x - k) \tag{43}$$

is solved by the father function, called φ, and the mother function ψ is given by

$$\psi(x) = \sum_{k \in \mathbb{Z}} b_k \varphi(2x - k). \tag{44}$$

A fundamental question is the converse one: Give simple conditions on two sequences (a_k) and (b_k) which guarantee the existence of $L^2(\mathbb{R})$-solutions φ and ψ which satisfy the orthogonality relations for the translates (41). How do we then get an orthogonal basis from this? The identities for Haar's functions φ and ψ of (40)(a) and (40)(b) above make it clear that the answer lies in a similar tiling and matching game which is implicit in the more general identities (43) and (44). Clearly we might ask the same question for other scaling numbers, for example $x \to 3x$ or $x \to 4x$ in place of $x \to 2x$. Actually a direct analogue of the visual interpretation from (40) makes it clear that there are no nonzero locally integrable solutions to the simple variants of (43),

$$\varphi(x) = \frac{3}{2}(\varphi(3x) + \varphi(3x - 2)) \tag{45}$$

or

$$\varphi(x) = 2(\varphi(4x) + \varphi(4x - 2)). \tag{46}$$

There *are* nontrivial solutions to (45) and (46), to be sure, but they are versions of the Cantor Devil's Staircase functions, which are prototypes of functions which are not locally integrable.

Since the Haar example is based on the fitting of copies of a fixed "box" inside an expanded one, it would almost seem unlikely that the system (43)–(44) admits finite sequences (a_k) and (b_k) such that the corresponding solutions φ and ψ are continuous or differentiable functions of compact support. The discovery in the mid-1980's of compactly supported differentiable solutions, see [15], was paralleled by applications in seismology, acoustics [24], and optics [61], as discussed in [62], and once the solutions were found, other applications followed at a rapid pace: see, for example, the ten books in Benedetto's review [6]. It is the solution ψ in (44) that the fuss is about, the mother function; the other one, φ, the father function, is only there before the birth of the wavelet. The most famous of them are named after Daubechies, and look like the graphs in Figure 1. With the multiresolution idea, we arrive at the closed subspaces

$$\mathcal{V}_j := U^{-j}\mathcal{V}_0, \qquad j \in \mathbb{Z}, \tag{47}$$

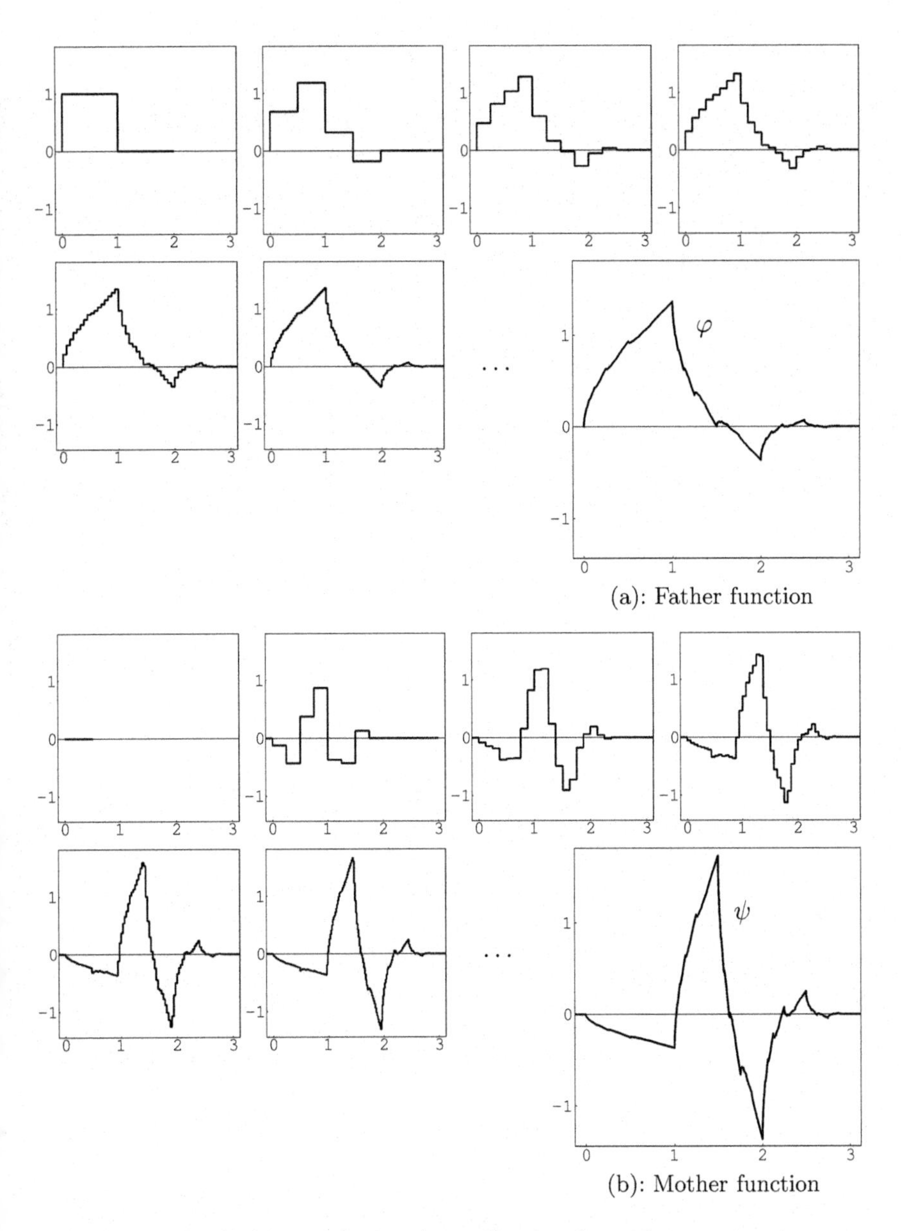

Fig. 1. Daubechies wavelet functions and series of cascade approximants.

as noted in (41)–(42), where U is some scaling operator. There are extremely effective iterative algorithms for solving the scaling identity (43): see, for example, Example 2.5.3, pp. 124–125, of [9]*, [15], and [80], and Figure 1. A key step in the algorithms involves a clever choice of the kind of resolution pictured in (52), but digitally encoded. The orthogonality relations can be encoded in the numbers (a_k) and (b_k) of (43)–(44), and we arrive at the doubly indexed functions

$$\psi_{j,k}(x) := 2^{j/2}\psi\left(2^j x - k\right), \qquad j, k \in \mathbb{Z}. \tag{48}$$

It is then not difficult to establish the combined orthogonality relations

$$\int_{\mathbb{R}} \overline{\psi_{j,k}(x)}\, \psi_{j',k'}(x)\ dx = \left\langle \psi_{j,k} \mid \psi_{j',k'} \right\rangle = \delta_{j,j'}\delta_{k,k'} \tag{49}$$

plus the fact that the functions in (48) form an orthogonal basis for $L^2(\mathbb{R})$. This provides a painless representation of $L^2(\mathbb{R})$-functions

$$f = \sum_{j\in\mathbb{Z}} \sum_{k\in\mathbb{Z}} c_{j,k}\psi_{j,k} \tag{50}$$

where the coefficients $c_{j,k}$ are

$$c_{j,k} = \int_{\mathbb{R}} \overline{\psi_{j,k}(x)}\, f(x)\ dx = \left\langle \psi_{j,k} \mid f \right\rangle. \tag{51}$$

What is more significant is that the resolution structure of closed subspaces of $L^2(\mathbb{R})$

$$\cdots \subset \mathcal{V}_{-2} \subset \mathcal{V}_{-1} \subset \mathcal{V}_0 \subset \mathcal{V}_1 \subset \mathcal{V}_2 \subset \cdots \tag{52}$$

facilitates powerful algorithms for the representation of the numbers $c_{j,k}$ in (51). Amazingly, the two sets of numbers (a_k) and (b_k) which were used in (43)–(44), and which produced the magic basis (48), the wavelets, are the same magic numbers which encode the quadrature mirror filters of signal processing of communications engineering. On the face of it, those signals from communications engineering really seem to be quite unrelated to the issues from wavelets—the signals are just sequences, time is discrete, while wavelets concern $L^2(\mathbb{R})$ and problems in mathematical analysis that are highly non-discrete. Dual filters, or more generally, subband filters, were invented in engineering well before the wavelet craze in mathematics of

*See an implementation of the "cascade" algorithm using Mathematica, and a "cartoon" of wavelets computed with it, at
http://www.math.uiowa.edu/~jorgen/wavelet_motions.pdf .

recent decades. These dual filters in engineering have long been used in technology, even more generally than merely for the context of quadrature mirror filters (QMF's), and it turns out that other popular dual wavelet bases for $L^2(\mathbb{R})$ can be constructed from the more general filter systems; but the best of the wavelet bases are the ones that yield the strongest form of orthogonality, which is (49), and they are the ones that come from the QMF's. The QMF's in turn are the ones that yield perfect reconstruction of signals that are passed through filters of the analysis-synthesis algorithms of signal processing. They are also the algorithms whose iteration corresponds to the resolution system (52) from wavelet theory.

While Fourier invented his transform for the purpose of solving the heat equation, i.e., the partial differential equation for heat conduction, the wavelet transform (50)–(51) does not diagonalize the differential operators in the same way. Its effectiveness is more at the level of computation; it turns integral operators into sparse matrices, i.e., matrices which have "many" zeros in the off-diagonal entry slots. Again, the resolution (52) is key to how this matrix encoding is done in practice.

1.2.1. *Some points of history*

The first wavelet was discovered by Alfred Haar long ago, but its use was limited since it was based on step-functions, and the step-functions jump from one step to the next. The implementation of Haar's wavelet in the approximation problem for continuous functions was therefore rather bad, and for differentiable functions it is atrocious, and so Haar's method was forgotten for many years. And yet it had in it the one idea which proved so powerful in the recent rebirth (since the 1980's) of wavelet analysis: the idea of a *multiresolution*. You see it in its simplest form by noticing that a box function B of (53) may be scaled down by a half such that two copies B' and B'' of the smaller box then fit precisely inside B. See (53).

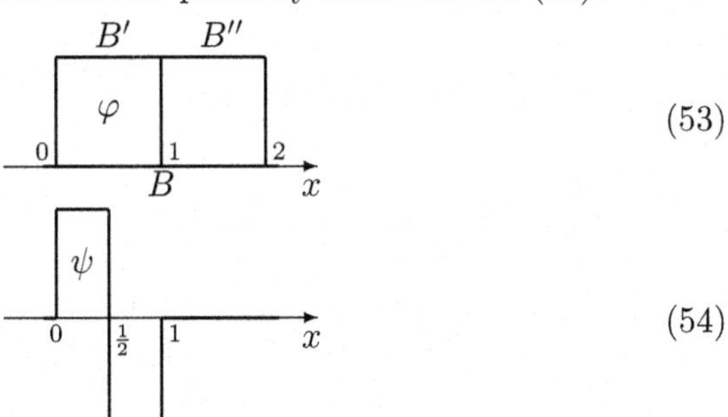

This process may be continued if you scale by powers of 2 in both directions, i.e., by 2^k for integral k, $-\infty < k < \infty$. So for every $k \in \mathbb{Z}$, there is a finer resolution, and if you take an up- and a shifted mirror image down-version of the dyadic scaling as in (54), and allow all linear combinations, you will notice that arbitrary functions f on the line $-\infty < x < \infty$, with reasonable integrability properties, admit a representation

$$f(x) = \sum_{k,n} c_{k,n} \psi\left(2^k x - n\right), \tag{55}$$

where the summation is over all pairs of integers $k, n \in \mathbb{Z}$, with k representing scaling and n translation. The very simple idea of turning this construction into a multiresolution ("multi" for the variety of scales in (55)) leads not only to an algorithm for the analysis/synthesis problem,

$$f(x) \longleftrightarrow c_{k,n}, \tag{56}$$

in (55), but also to a construction of the single functions ψ which solve the problem in (55), and which can be chosen differentiable, and yet with support contained in a fixed finite interval. These two features, the algorithm and the finite support (called *compact* support), are crucial for computations: Computers do algorithms, but they do not do infinite intervals well. Computers do summations and algebra well, but they do not do integrals and differential equations, unless the calculus problems are discretized and turned into algorithms.

In the discussion to follow, the multiresolution analysis viewpoint is dominant, which increases the role of algorithms; for example, the so-called pyramid algorithm for analyzing signals, or shapes, using wavelets, is an outgrowth of multiresolutions.

Returning to (53) and (54), we see that the scaling function φ itself may be expanded in the wavelet basis which is defined from ψ, and we arrive at the infinite series

$$\varphi(x) = \sum_{k=1}^{\infty} 2^{-k} \psi\left(2^{-k} x\right) \tag{57}$$

which is pointwise convergent for $x \in \mathbb{R}$. (It is a special case of the expansion (55) when $f = \varphi$.) In view of the picture (⊳) below, (57) gives an alternative meaning to the traditional concept of a *telescoping* infinite sum. If, for example, $0 < x < 1$, then the representation (57) yields $\varphi(x) = 1 = \frac{1}{2} + \left(\frac{1}{2}\right)^2 + \cdots$, while for $1 < x < 2$, $\varphi(x) = 0 = -\frac{1}{2} + \left(\frac{1}{2}\right)^2 + \left(\frac{1}{2}\right)^3 + \cdots$.

More generally, if $n \in \mathbb{N}$, and $2^{n-1} < x < 2^n$, then

$$\varphi(x) = 0 = -\left(\frac{1}{2}\right)^n + \sum_{k>n} \left(\frac{1}{2}\right)^k .$$

So the function φ is itself in the space $\mathcal{V}_0 \subset L^2(\mathbb{R})$, and φ represents the *initial resolution.* The tail terms in (57) corresponding to

$$\sum_{k>n} 2^{-k}\psi\left(2^{-k}x\right) = \frac{1}{2^n}\varphi\left(\frac{x}{2^n}\right) \tag{58}$$

represent the *coarser resolution.* The finite sum

$$\sum_{k=1}^{n} 2^{-k}\psi\left(2^{-k}x\right)$$

represents the *missing detail* of φ as a "bump signal". While the sum on the left-hand side in (58) is *infinite*, i.e., the summation index k is in the range $n < k < \infty$, the expression $2^{-n}\varphi\left(2^{-n}x\right)$ on the right-hand side is merely a coarser scaled version of the original function φ from the subspace $\mathcal{V} \subset L^2(\mathbb{R})$ which specifies the initial resolution. Infinite sums are *analysis problems* while a scale operation is a single simple *algorithmic step.* And so we have encountered a first (easy) instance of the magic of a resolution algorithm; i.e., an instance of a transcendental step (the analysis problem) which is converted into a programmable operation, here the operation of scaling. (Other more powerful uses of the scaling operation may be found in the recent book [63] by Yves Meyer, especially Ch. 5, and [38].)

The sketch below allows you to visualize more clearly this resolution versus detail concept which is so central to the wavelet algorithms, also for general wavelets which otherwise may be computationally more difficult than the Haar wavelet.

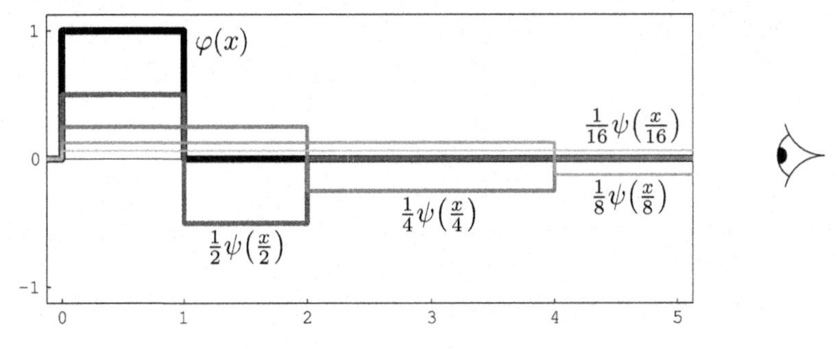

The wavelet decomposition of Haar's bump function φ in (53) and (57)

Using the sketch we see for example that the simple step function

$$f(x) = a\varphi(x) + b\varphi(x-1) = a\chi_{[0,1\rangle}(x) + b\chi_{[1,2\rangle}(x) \tag{59}$$

has the wavelet decomposition into a sum of a *coarser resolution* and an *intermediate detail* as follows:

$$f(x) = \underbrace{\frac{a-b}{2}\psi\left(\frac{x}{2}\right)}_{\text{intermediate detail}} + \underbrace{\frac{a+b}{2}\varphi\left(\frac{x}{2}\right)}_{\text{coarser version}}, \qquad x \in \mathbb{R}. \tag{60}$$

Thus the details are measured as differences. This is a general feature that is valid for other functions and other wavelet resolutions. See, for instance, § 2.2 below.

1.2.2. *Some early applications*

While the Haar wavelet is built from flat pieces, and the orthogonality properties amount to a visual tiling of the graphs of the two functions φ and ψ, this is not so for the Daubechies wavelet nor the other compactly supported smooth wavelets. By the Balian–Low theorem [15], a time-frequency wavelet cannot be simultaneously localized in the two dual variables: if ψ is a time-frequency Gabor wavelet, then the two quantities $\int_{\mathbb{R}} |x\psi(x)|^2\, dx$ and $\int_{\mathbb{R}} \left|t\hat{\psi}(t)\right|^2 dt$ cannot both be finite. Since $\widehat{\left(\frac{d\psi}{dx}\right)}(t) = it\hat{\psi}(t)$, this amounts to poor differentiability properties of well-localized Gabor wavelets, i.e., wavelets built using the two operations translation and frequency modulation over a lattice.

But with the multiresolution viewpoint, we can understand the first of Daubechies's scaling functions as a one-sided differentiable solution φ to

$$\varphi(x) = h_0\varphi(2x) + h_1\varphi(2x-1) + h_2\varphi(2x-2) + h_3\varphi(2x-3), \tag{61}$$

where the four real coefficients satisfy

$$\left.\begin{aligned} h_0 + h_1 + h_2 + h_3 &= 2, \\ h_3 - h_2 + h_1 - h_0 &= 0, \\ h_3 - 2h_2 + 3h_1 - 4h_0 &= 0, \\ h_1 h_3 + h_0 h_2 &= 0. \end{aligned}\right\} \tag{62}$$

The system (62) is easily solved:

$$\left.\begin{aligned} 4h_0 = 1 + \sqrt{3}, \quad 4h_2 = 3 - \sqrt{3}, \\ 4h_1 = 3 + \sqrt{3}, \quad 4h_3 = 1 - \sqrt{3}, \end{aligned}\right\} \tag{63}$$

and Daubechies showed that (61) has a solution φ which is supported in the interval $[0, 3]$, is one-sided differentiable, and satisfies the conditions

$$\int_{\mathbb{R}} \varphi(x)\, dx = 1, \quad \int_{\mathbb{R}} \psi(x)\, dx = 0, \quad \text{and} \quad \int_{\mathbb{R}} x\psi(x)\, dx = 0. \tag{64}$$

The first applications served as motivating ideas as well: optics, seismic measurements, dynamics, turbulence, data compression; see the book [54] Actually, it is two books: the first one (primarily by Kahane) is classical Fourier analysis, and the second one (primarily by P.-G. Lemarié-Rieusset) is the wavelet book. It will help you, among other things, to get a better feel for the French connection, the Belgian connection, and the diverse and early impulses from applications in the subject. Enjoy!

For a list of more recent applications we recommend [64].

2. Signal Processing

If we idealize and view time as discrete, a copy of $\mathbb{Z}$, then a signal is a sequence $(\xi_n)_{n\in\mathbb{Z}}$ of numbers. A filter is an operator which calculates weighted averages

$$(\xi_n) \longmapsto \sum_{k\in\mathbb{Z}} a_k \xi_{n-k}. \tag{65}$$

But working instead with functions of $z \in \mathbb{T}$, this is multiplication, $f(z) \mapsto m(z) f(z)$, where $m(z) = \sum_{k\in\mathbb{Z}} a_k z^k$ and $f(z) = \sum_{k\in\mathbb{Z}} \xi_k z^k$ are the usual Fourier representation of the corresponding generating functions. Similarly, down-sampling $(N\!\downarrow)$ and up-sampling $(N\!\uparrow)$ as operators on sequences take the form

$$f \longmapsto \frac{1}{N} \sum_{w\in\mathbb{T},\, w^N = z} f(w) \tag{66}$$

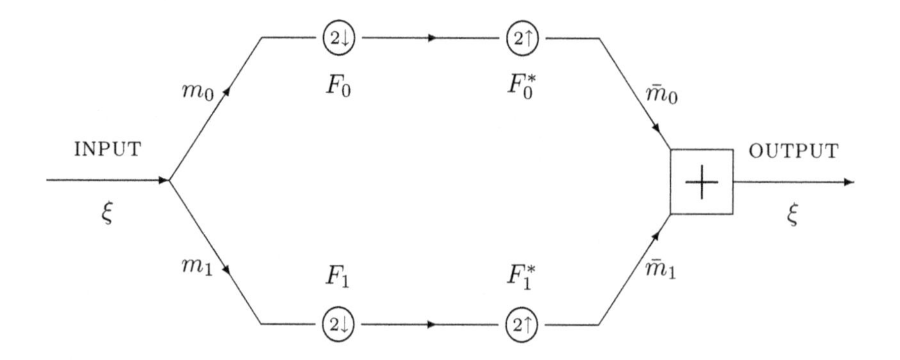

Fig. 2. Perfect reconstruction of signals.

and

$$f \longmapsto f\left(z^N\right). \tag{67}$$

Since the operators $(N\downarrow)$ and $(N\uparrow)$ are clearly dual to one another on the Hilbert space $\ell^2(\mathbb{Z})$ of sequences (i.e., time-signals), we get the corresponding duality for $L^2(\mathbb{T})$, i.e.,

$$\int_{\mathbb{T}} f\left(z^N\right) g(z)\, d\mu(z) = \int_{\mathbb{T}} f(z) \frac{1}{N} \sum_{w^N = z} g(w)\, d\mu(z), \tag{68}$$

where μ denotes the normalized Haar measure on $\mathbb{T}$, or equivalently the following identity for 2π-periodic functions:

$$\int_0^{2\pi} f(N\theta)\, g(\theta)\, d\theta = \int_0^{2\pi} f(\theta) \frac{1}{N} \sum_{k=0}^{N-1} g\left(\frac{\theta + k \cdot 2\pi}{N}\right) d\theta. \tag{69}$$

Quadrature mirror filters with N frequency subbands $m_0, m_1, \dots, m_{N-1}$ give perfect reconstruction when signals are analyzed into subbands and then reconstructed via the up-sampling and corresponding dual filters. In engineering formalism this is expressed in the diagram in Fig. 2, for $N = 2$, and m_0, resp. m_1, are called low-pass, resp. high-pass, filters. In operator language, this takes the form

$$F_0^* F_0 + F_1^* F_1 = I,$$

where F_0 and F_1 are the operators in Fig. 2, with dual operators F_0^* and F_1^*. The quadrature conditions may be expressed as

$$F_0 F_0^* = F_1 F_1^* = I \tag{70}$$

and

$$F_0 F_1^* = F_1 F_0^* = 0. \tag{71}$$

In operator theory there is tradition for working instead with the operators $S_j := F_j^*$. When viewed as operators on $L^2(\mathbb{T})$ they are therefore isometries with orthogonal ranges, and they satisfy

$$\sum_{j=0}^{N-1} S_j S_j^* = I \tag{72}$$

with I now representing the identity operator acting on $L^2(\mathbb{T})$. The relations on the S_j-operators are known as the Cuntz relations because of their use in C^*-algebra theory; see [13]. In the present application they take the form

$$(S_j f)(z) = m_j(z) f(z^N), \qquad f \in L^2(\mathbb{T}), \tag{73}$$

and

$$(S_j^* f)(z) = \frac{1}{N} \sum_{w^N = z} \overline{m_j(w)} f(w), \tag{74}$$

and the Cuntz relations are equivalent to the conditions

$$\sum_{w^N = z} |m_j(w)|^2 = N \tag{75}$$

and

$$\sum_{w^N = z} \overline{m_j(w)} m_k(w) = 0 \qquad \text{for all } z \in \mathbb{T} \text{ and } j \neq k. \tag{76}$$

The last conditions are known in engineering as the quadrature conditions for the subband filters $m_0, m_1, \ldots, m_{N-1}$, with m_0 denoting the low-pass filter. The low-pass and band-pass conditions on the functions m_j are perhaps more familiar in the additive notation given by the substitution $z := e^{-i\theta}$. Then the functions m_j are viewed as 2π-periodic, and

$$m_j\left(j \cdot \frac{2\pi}{N}\right) = \sqrt{N},$$

while

$$m_j\left(k\cdot\frac{2\pi}{N}\right)=0 \qquad \text{for } j\neq k,$$

with both of the indices j, k ranging over $0,1,\ldots,N-1$.

2.1. *Filters in communications engineering*

The coefficients of the functions $m_j(\cdot)$ are called *impulse response coefficients* in communications engineering, and when used in wavelets and in subdivision algorithms, they are called *masking coefficients.* In the finite case, the $m_j(\cdot)$'s are also called FIR for finite impulse response. The model illustrated in Fig. 2 is used in filter design in either hardware or software:

⟦1⟧ Try filters m_0, m_1 in Fig. 2, and approximate the output to the input;
⟦2⟧ Choose a specific structure in which the filter will be realized and then quantize the coefficients, length and numerical values;
⟦3⟧ Verify by simulation that the resulting design meets given performance specifications.

Once filters are constructed, we saw that they are also providing us with wavelet algorithms. When the steps of Fig. 2 are iterated, we arrive at wavelet subdivision algorithms. Relative to a given resolution (pictured as a closed subspace $\mathcal{V}_1$, say, in $L^2(\mathbb{R})$), signals, i.e., functions in $L^2(\mathbb{R})$, decompose into coarser ones and intermediate details. Relative to the subspaces $\mathcal{W}_0$ and $\mathcal{V}_1$, this amounts to

$$\begin{array}{ccccc} \mathcal{V}_1 & = & \mathcal{V}_0 & + & \mathcal{W}_0. \\ \uparrow & & \uparrow & & \uparrow \\ \text{given} & & \text{coarser} & & \text{intermediate} \\ \text{resolution} & & \text{resolution} & & \text{detail} \end{array} \tag{77}$$

Ideally, we wish the decomposition in (77) to be orthogonal in the sense that

$$\langle f \mid g\rangle = 0 \qquad \text{for all } f\in\mathcal{V}_0 \text{ and all } g\in\mathcal{W}_0. \tag{78}$$

Since the subdivisions involve translations by discrete steps, we specialize the resolution such that both of the spaces $\mathcal{V}_0$ and $\mathcal{W}_0$ are invariant under translations by points in $\mathbb{Z}$, i.e., such that

$$T\colon f\longmapsto f(\cdot-1) \tag{79}$$

leaves both of the subspaces $\mathcal{V}_0$ and $\mathcal{W}_0$ invariant. The *multiresolution analysis* case (MRA) corresponds to the setup when $\mathcal{V}_0$ is singly generated, i.e., there is a function $\varphi \in \mathcal{V}_0$ such that the closed linear span of

$$T_n\varphi(\cdot) = \varphi(\cdot - n), \qquad n \in \mathbb{Z}, \tag{80}$$

is all of $\mathcal{V}_0$. If $N = 2$, then there is also a $\psi \in \mathcal{W}_0$ such that the closed linear span of $\{\psi(\cdot - n) : n \in \mathbb{Z}\}$ is all of $\mathcal{W}_0$. If $N > 2$, we may need functions $\psi_1, \dots, \psi_{N-1}$ in $\mathcal{W}_0$ such that $\{\psi_i(\cdot - n) : i = 1, \dots, N-1,\ n \in \mathbb{Z}\}$ has a closed span equal to $\mathcal{W}_0$.

2.2. *Algorithms for signals and for wavelets*

The pyramid algorithm and the Cuntz relations. Since the two Hilbert spaces $L^2(\mathbb{T})$ and $\ell^2(\mathbb{Z})$ are isomorphic via the Fourier series representation, it follows that the system $\{S_i\}_{i=0}^1$ is equivalent to a system $\{\hat{S}_i\}_{i=0}^1$ acting on $\ell^2(\mathbb{Z})$. Specifically, $(S_i f)\widehat{\ } = \hat{S}_i \hat{f}$, $i = 0, 1$, where $\hat{f}(n) := \int_{\mathbb{T}} z^{-n} f(z)\, d\mu(z)$. For $c := (c_n)_{n\in\mathbb{Z}}$ in $\ell^2(\mathbb{Z})$, and functions f on $\mathbb{R}$, set

$$f_{-1}(x) := (Uf)(x) = 2^{-\frac{1}{2}} f\left(\frac{x}{2}\right), \text{ and}$$
$$(c * f)(x) := \sum_{n\in\mathbb{Z}} c_n f(x-n).$$

For the present, let $\{m_i\}_{i=0}^1$ be the low-pass and high-pass wavelet filters, and let φ, ψ be the corresponding scaling function, resp., wavelet function, also called father function, resp., mother function. Now introduce the corresponding operators S_i and their cousins $\hat{S}_i$. The adjoints $\hat{S}_i^*$ are also called *filters*.

Then

$$c * \varphi = \underbrace{\left(\left(\hat{S}_0^* c\right) * \varphi\right)_{-1}}_{\text{coarser resolution}} + \underbrace{\left(\left(\hat{S}_1^* c\right) * \psi\right)_{-1}}_{\text{detail}} \qquad \text{for all } c \in \ell^2(\mathbb{Z}). \tag{81}$$

Define $W : \ell^2 \to \ell^2$ by

$$W(c)(x) = (c * \varphi)(x) = \sum_{n\in\mathbb{Z}} c_n \varphi(x-n). \tag{82}$$

Then W maps ℓ^2 isometrically onto $\mathcal{V}_0$ in the orthogonal case and

$$W\hat{S}_0 = UW.$$

Further

$$W\hat{S}_0\hat{S}_0^*c = \left(\hat{S}_0^*c * \varphi\right)_{-1}.$$

Embedding ℓ^2 into $\ell^2 \oplus \ell^2$ as $\ell^2 \oplus 0$, extend W to $\ell^2 \oplus \ell^2$ by putting

$$W(c \oplus d) = c * \varphi + d * \psi.$$

Then the extended W maps $\ell^2 \oplus \ell^2$ isometrically onto $U^{-1}\mathcal{V}_0$ and

$$W\left(\hat{S}_0 c + \hat{S}_1 d\right) = UW(c \oplus d)$$

for all $c, d \in \ell^2$, where the left W is the one from (82) and the right is the extension of W to $\ell^2 \oplus \ell^2$.

At this point you can use $1_{\ell^2} = \hat{S}_0\hat{S}_0^* + \hat{S}_1\hat{S}_1^*$ to show (81). Note that if $c_0 = a$ and $c_1 = b$ and $c_i = 0$ for other i, the formula (81) reduces to (60).

The subdivision relations (81) are equivalent to the system

$$\sqrt{2}\varphi(2x) = \sum_{k\in\mathbb{Z}} \bar{a}_{2k}\varphi(x+k) + \sum_{k\in\mathbb{Z}} \bar{b}_{2k}\psi(x+k), \tag{83}$$

$$\sqrt{2}\varphi(2x-1) = \sum_{k\in\mathbb{Z}} \bar{a}_{2k+1}\varphi(x+k) + \sum_{k\in\mathbb{Z}} \bar{b}_{2k+1}\psi(x+k), \tag{84}$$

where the coefficients a_n, b_n are those of the quantum wavelet algorithm, i.e., the coefficients in the "large" unitary matrix (85). Thus the quantum algorithm does the wavelet decomposition within a fixed resolution subspace.

The scaling function φ defines a resolution subspace $\mathcal{V}_0 \subset L^2(\mathbb{R})$. Then (81), or equivalently (83)–(84), represents the orthogonal decomposition of functions in $\mathcal{V}_0$ into an orthogonal sum of a function with coarser resolution and a function in the intermediate detail subspace.

Let m_0, m_1 be a dyadic wavelet filter, and let $\mathbb{T} \ni z \mapsto A(z) \in \mathrm{U}_2(\mathbb{C})$ be the corresponding matrix function, $A_{i,j}(z) = \frac{1}{2}\sum_{w^2=z} w^{-j}m_i(w)$. If the low-pass filter $m_0(z) = a_0 + a_1 z + \cdots + a_{2n+1}z^{2n+1}$, then a choice for $m_1(z) = \sum_{k=0}^{2n+1} b_k z^k$ is $b_k = (-1)^k \bar{a}_{2n+1-k}$. We then have $A(z) = \sum_{k=0}^{n} A_k z^k$ where $A_k = \begin{pmatrix} a_{2k} & a_{2k+1} \\ b_{2k} & b_{2k+1} \end{pmatrix}$, and the following $2^{n+2} \times 2^{n+2}$ scalar matrix can be checked to be unitary:

$$
\left(\begin{array}{c|c|c|c|c|c|c|c|c|c|c|c}
\begin{matrix}a_1\\ b_1\end{matrix} & A_1 & A_2 & \cdots & A_{n-1} & A_n & 0 & & \cdots & & 0 & \begin{matrix}a_0\\ b_0\end{matrix} \\ \hline
\begin{matrix}0\\ 0\end{matrix} & A_0 & A_1 & \cdots & A_{n-2} & A_{n-1} & A_n & 0 & & \cdots & 0 & \begin{matrix}0\\ 0\end{matrix} \\ \hline
\begin{matrix}0\\ 0\end{matrix} & 0 & A_0 & \cdots & A_{n-3} & A_{n-2} & A_{n-1} & A_n & 0 & \cdots & 0 & \begin{matrix}0\\ 0\end{matrix} \\ \hline
\begin{matrix}0\\ 0\end{matrix} & & & & & & & & & & & \begin{matrix}0\\ 0\end{matrix} \\ \hline
\vdots & & & & \ddots & & & & & \ddots & & \vdots \\ \hline
\begin{matrix}0\\ 0\end{matrix} & & & & & & & & & & & \begin{matrix}0\\ 0\end{matrix} \\ \hline
\begin{matrix}a_{2n+1}\\ b_{2n+1}\end{matrix} & 0 & & \cdots & & 0 & A_0 & A_1 & A_2 & \cdots & A_{n-1} & \begin{matrix}a_{2n}\\ b_{2n}\end{matrix} \\ \hline
\begin{matrix}a_{2n-1}\\ b_{2n-1}\end{matrix} & A_n & 0 & & \cdots & & 0 & A_0 & A_1 & \cdots & A_{n-2} & \begin{matrix}a_{2n-2}\\ b_{2n-2}\end{matrix} \\ \hline
\begin{matrix}a_{2n-3}\\ b_{2n-3}\end{matrix} & A_{n-1} & A_n & 0 & & \cdots & & 0 & A_0 & \cdots & A_{n-3} & \begin{matrix}a_{2n-4}\\ b_{2n-4}\end{matrix} \\ \hline
\vdots & & & \ddots & & & & & & \ddots & & \vdots \\ \hline
\begin{matrix}a_3\\ b_3\end{matrix} & A_2 & A_3 & \cdots & A_n & 0 & & \cdots & & 0 & A_0 & \begin{matrix}a_2\\ b_2\end{matrix}
\end{array}\right) \tag{85}
$$

Except for the scalar entries in the two extreme left and right columns, all the other entries of the big combined matrix U_A are taken from the cyclic arrangements of the 2×2 matrices of coefficients $A_0, A_1, \ldots, A_n$ in the expansion of $A(z)$. For the case of $n = 1$ this amounts to the simple 8×8 wavelet matrix

$$
A_0 \circlearrowleft \quad
\left(\begin{array}{c|c|c|c|c}
\begin{matrix}a_1\\ b_1\end{matrix} & A_1 & 0 & 0 & \begin{matrix}a_0\\ b_0\end{matrix} \\ \hline
\begin{matrix}0\\ 0\end{matrix} & A_0 & A_1 & 0 & \begin{matrix}0\\ 0\end{matrix} \\ \hline
\begin{matrix}0\\ 0\end{matrix} & 0 & A_0 & A_1 & \begin{matrix}0\\ 0\end{matrix} \\ \hline
\begin{matrix}a_3\\ b_3\end{matrix} & 0 & 0 & A_0 & \begin{matrix}a_2\\ b_2\end{matrix}
\end{array}\right)
\quad \circlearrowright A_1, \tag{86}
$$

which is the one that produces the sequence of quantum gates. The quantum algorithm of a wavelet filter is thus represented by a $2^{n+2} \times 2^{n+2}$ unitary matrix U_A acting on the quantum qubit register $\underbrace{\mathbb{C} \otimes \cdots \otimes \mathbb{C}}_{n+2 \text{ times}} = \mathbb{C}^{2(n+2)}$, i.e., it acts on a configuration of $n+2$ qubits. The realization of a wavelet algorithm in the quantum realm thus amounts to spelling out the steps in factoring U_A into a product of qubit gates. By Shor's theorem, we know that this can be done, and U_A may be built out of one-qubit gates and CNOT gates following the ideas sketched above. The reader may find more discussion of the matrix U_A in Section 3 of [26].

The generalization of classical and quantum wavelet resolution algorithms from $N = 2$ to $N > 2$ is immediate: Then $m_i(z) = \sum_{k\in\mathbb{Z}} a_k^{(i)} z^k$,

$$(S_i f)(z) = m_i(z) f(z^N), \qquad i = 0, \dots, N-1, \tag{87}$$

and the transformation rules

$$\xi_{Nk+i} = \sum_{l\in\mathbb{Z}} a_{l-Nk}^{(i)} \varepsilon_l, \qquad i = 0, 1, \dots, N-1, \tag{88}$$

permute the set of ONB's in $\ell^2(\mathbb{Z})$ and define a unitary commuting with the N-shift. Hence, the standard formulas from [88], [56], and [25] for the quantum computing algorithm naturally generalize to the case $N > 2$ via (88). Instead of k-registers $\underbrace{\mathbb{C}^2 \otimes \cdots \otimes \mathbb{C}^2}_{k \text{ times}} = \mathbb{C}^{2^k}$ over $\mathbb{C}^2$, we will now have to work rather with $\underbrace{\mathbb{C}^N \otimes \cdots \otimes \mathbb{C}^N}_{k \text{ times}} = \mathbb{C}^{N^k}$.

The use of the algorithmic relations in engineering and operator algebra theory predates their more recent use in wavelet theory and wavepacket analysis.

2.2.1. *Pyramid algorithms*

For $N > 2$, the algorithm of the previous section takes the following form.

The pyramid algorithm and the Cuntz relations revisited. By Fourier equivalence of $L^2(\mathbb{T})$ and $\ell^2(\mathbb{Z})$ via the Fourier series, it follows that the system $\{S_i\}_{i=0}^{N-1}$ is equivalent to a system $\{\hat{S}_i\}_{i=0}^{N-1}$ acting on $\ell^2(\mathbb{Z})$. Specifically, $(S_i f)\hat{} = \hat{S}_i \hat{f}$, $i = 0, \dots, N-1$, where $\hat{f}(n) := \int_{\mathbb{T}} z^{-n} f(z)\, d\mu(z)$. For $c := (c_n)_{n\in\mathbb{Z}}$ in $\ell^2(\mathbb{Z})$, and functions f on $\mathbb{R}$, set

$$f_{-1}(x) := N^{-\frac{1}{2}} f\left(\frac{x}{N}\right),$$

and

$$(c * f)(x) := \sum_{n \in \mathbb{Z}} c_n f(x - n).$$

Let $\{m_i\}_{i=0}^{N-1}$ be low-pass and high-pass wavelet filters, and let φ, $\psi_1, \ldots, \psi_{N-1}$ be the corresponding scaling function, resp., wavelet functions. Now introduce the corresponding operators S_i, and their cousins $\hat{S}_i$. The adjoints $\hat{S}_i^*$ are also called *filters.*

Then

$$c * \varphi = \underbrace{\left(\left(\hat{S}_0^* c\right) * \varphi\right)_{-1}}_{\text{coarser resolution}} + \underbrace{\sum_{i=1}^{N-1} \left(\left(\hat{S}_i^* c\right) * \psi_i\right)_{-1}}_{\text{detail}} \qquad \text{for all } c \in \ell^2(\mathbb{Z}). \quad (89)$$

The scaling function φ defines a resolution subspace $\mathcal{V}_0 \subset L^2(\mathbb{R})$. For the case $N > 2$:

Discrete vs. continuous wavelets, i.e., ℓ^2 vs. $L^2(\mathbb{R})$:

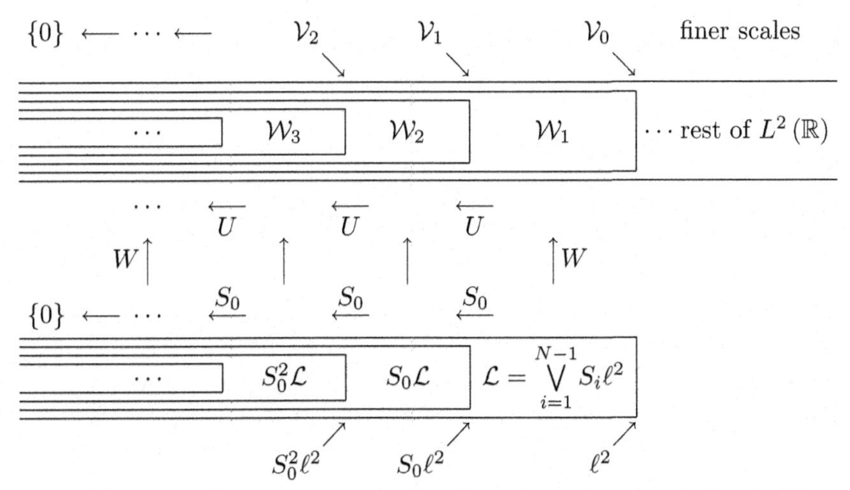

More refined pyramid algorithms yield wavelet packets as follows.

The Haar wavelet is supported in $[0, 1]$, and if $j \in \mathbb{Z}_+$ and $k \in \mathbb{Z}$, then the modified function $x \mapsto \psi\left(2^j x - k\right)$ is supported in the smaller interval $\frac{k}{2^j} \leq x \leq \frac{k+1}{2^j}$. When j is fixed, these intervals are contained in $[0, 1]$ for $k \in \left\{0, 1, \ldots, 2^j - 1\right\}$. This is not the case for the other wavelet functions. For one thing, the non-Haar wavelets ψ have support intervals of length more than one, and this forces periodicity considerations; see [11]. For this reason, Coifman and Wickerhauser [12] invented the concept of wavelet packets. They are built from functions with prescribed smoothness, and yet they have localization properties that rival those of the (discontinuous) Haar wavelet.

There are powerful but nontrivial theorems on restriction algorithms for wavelets $\psi_{j,k}(x) = 2^{\frac{j}{2}}\psi\left(2^j x - k\right)$ from $L^2(\mathbb{R})$ to $L^2(0,1)$. We refer the reader to [11] and [65] for the details of this construction. The underlying idea of Alfred Haar has found a recent renaissance in the work of Wickerhauser [88] on *wavelet packets.* The idea there, which is also motivated by the Walsh function algorithm, is to replace the refinement equation (33) by a related recursive system as follows: Let $m_0(z) = \sum_k a_k z^k$, $m_1(z) = \sum_k b_k z^k$, for example $b_k = (-1)^k \bar{a}_{1-k}$, $k \in \mathbb{Z}$, be a given low-pass/high-pass system, $N = 2$. Then consider the following *refinement system* on $\mathbb{R}$:

$$W_{2n}(x) = \sqrt{2}\sum_{k\in\mathbb{Z}} a_k W_n(2x-k), \tag{90}$$

$$W_{2n+1}(x) = \sqrt{2}\sum_{k\in\mathbb{Z}} b_k W_n(2x-k). \tag{91}$$

Clearly the function W_0 can be identified with the traditional scaling function φ of (107). A theorem of Coifman and Wickerhauser (Theorem 8.1, [12]) states that if $\mathcal{P}$ is a partition of $\{0,1,2,\dots\}$ into subsets of the form

$$I_{k,n} = \left\{2^k n, 2^k n + 1, \dots, 2^k(n+1) - 1\right\},$$

then the function system

$$\left\{2^{\frac{k}{2}} W_n\left(2^k x - l\right) \,\middle|\, I_{k,n} \in \mathcal{P},\, l \in \mathbb{Z}\right\}$$

is an orthonormal basis for $L^2(\mathbb{R})$. Although it is not spelled out in [12], this construction of bases in $L^2(\mathbb{R})$ divides itself into the two cases, the true orthonormal basis (ONB), and the weaker property of forming a function system which is only a tight frame. As in the wavelet case, to get the $\mathcal{P}$-system to really be an ONB for $L^2(\mathbb{R})$, we must assume the transfer operator $R_{|m_0|^2}$ to have *Perron–Frobenius spectrum* on $C(\mathbb{T})$. This means that the intersection of the point spectrum of $R_{|m_0|^2}$ with $\mathbb{T}$ is the singleton $\lambda = 1$, and that $\dim\ker((1 - R_{|m_0|^2})|_{C(\mathbb{T})}) = 1$.

2.2.2. *Subdivision algorithms*

The algorithms for wavelets and wavelet packets involve the pyramid idea as well as subdivision. Each subdivision produces a multiplication of subdivision points. If the scaling is by N, then j subdivisions multiply the number of subdivision points by N^j. If the scaling is by a $d \times d$ integral matrix $\mathbf{N}$, then the multiplicative factor is $|\det \mathbf{N}|^j$ in the number of subdivision points placed in $\mathbb{R}^d$.

In the discussion below, we restrict attention to $d = 1$, but the conclusions hold with only minor modification in the general case of $d > 1$ and matrix scaling.

If W is a continuous function on $\mathbb{T}$, the *transfer operator* or *kneading operator* R_W

$$R_W \xi (z) = \frac{1}{N} \sum_{w^N = z} W(w) \xi(w) = S_0^* W \xi(z), \tag{92}$$

with the alias

$$(R_W f)_n = \sum_k c_{Nn-k} f_k \tag{93}$$

in the Fourier transformed space, has an adjoint which is the *subdivision operator* or chopping operator

$$(R_W^* \xi)(z) = \overline{W(z)} \xi(z^N) \tag{94}$$

on functions ξ on $\mathbb{T}$, with the alias

$$(R_W^* f)_n = \sum_k \overline{c_{Nk-n}} f_k \tag{95}$$

on sequences.

We will analyze the duality between R_W and R_W^* and their spectra. Specializing to $W = |m_0|^2$, we note that R_W is then the transfer operator of orthogonal type wavelets. In the following, W is assumed only to satisfy $W \in \mathrm{Lip}_1(\mathbb{T})$ and $W \geq 0$. Other conditions are discussed in [9].

In the engineering terminology of § 2.2, the operation (93) is composed of a local filter with the numbers c_k as coefficients, followed by the down-sampling $\textcircled{N\downarrow}$, while (95) is composed of up-sampling $\textcircled{N\uparrow}$, followed by an application of a dual filter. In signal processing, $\textcircled{N\downarrow}$ is referred to as "decimation" even if N is not 10.

The operator S $(= R_W^*)$ is called the subdivision operator, or the *woodcutter operator*, because of its use in computer graphics. Iterations of S will generate a shape which (in the case of one real dimension) takes the form of the graph of a function f on $\mathbb{R}$. If $\xi \in \ell^\infty(\mathbb{Z})$ is given, and if the differences

$$D_n(i) = f\left(\frac{i}{2^n}\right) - (S^n \xi)(i), \qquad i \in \mathbb{Z}, \tag{96}$$

are small, for example if

$$\lim_{n\to\infty} \sup_{i\in\mathbb{Z}} |D_n(i)| = 0, \tag{97}$$

then we say that ξ represents *control points*, or a control polygon, and the function f is the limit of the *subdivision scheme.*

It follows that the subdivision operator S on the sequence spaces, especially on $\ell^\infty(\mathbb{Z})$, governs *pointwise approximation* to refinable limit functions. The dual version of S, i.e., $R = S^*$ (= the transfer operator) governs the corresponding *mean approximation* problem, i.e., approximation relative to the $L^2(\mathbb{R})$-norm.

In Scholium 4.1.2 of [9], we consider the eigenvalue problem

$$S\xi = \lambda\xi, \qquad \lambda \in \mathbb{C}, \tag{98}$$

and $\xi \neq 0$ in some suitably defined space of sequences. The formula (96) for the limit of a given subdivision scheme S makes it clear that the case (98) must be excluded. For if (98) holds, for some $\lambda \in \mathbb{C}$, and some sequence ξ of control points, then there is not a corresponding regular function f on $\mathbb{R}$ with its values given on the finer grids $2^{-n}\mathbb{Z}$, $n = 1, 2, \ldots$, by

$$f_\xi\left(i2^{-n}\right) \approx (S^n\xi)(i) = \lambda^n\xi(i), \qquad i \in \mathbb{Z}. \tag{99}$$

We show in Example 4.1.3 of [9] that there are no such control points ξ in $\ell^2(\mathbb{Z}) \setminus \{0\}$. Hence the stability of the algorithm!

2.2.3. *Wavelet packet algorithms*

The main difference between the algorithms of wavelets and those of wavelet packets is that for the wavelets the path in the pyramid is to one side only: a given resolution is split into a coarser one and the intermediate detail. The intermediate detail may further be broken down into frequency bands. With the operators $S_j f(z) = m_j(z) f\left(z^N\right)$ acting on $L^2(\mathbb{T})$, the coarser subspace after j steps is modeled on $S_0^j L^2(\mathbb{T})$, and the projection onto this subspace is $S_0^j S_0^{*j}$ where S_0 is the isometry of $L^2(\mathbb{T}) \cong \mathcal{V}_0$ defined by the low-pass filter m_0. But in the construction of the wavelet packet, the subspace resulting by running the algorithm j times is $S_{i_1}S_{i_2}\cdots S_{i_j}L^2(\mathbb{T})$, and the projection onto this subspace is

$$S_{i_1}S_{i_2}\cdots S_{i_j}S_{i_j}^*\cdots S_{i_2}^*S_{i_1}^*.$$

If $n \in \mathbb{Z}_+$, the wavelet function W_n is computed from the iteration $i_1, \dots, i_j$ corresponding to the representation

$$n = i_1 + i_2 N + i_3 N^2 + \cdots + i_j N^{j-1},$$

where $i_1, \dots, i_j \in \{0, 1, \dots, N-1\}$ are unique from the Euclidean algorithm.

2.2.4. *Lifting algorithms: Sweldens and more*

The discussion centers around the matrix functions $A\colon \mathbb{T} \to \mathrm{GL}_2(\mathbb{C})$.

The case $\det A \equiv 1$. Recall that we call a finite sum $\sum_{k=-n_0}^{n_1} A_k z^k$, $n_0, n_1 \geq 0$, a Fourier polynomial both if the coefficients A_k are numbers, and if they are matrices. The matrix-valued Fourier polynomials $\mathbb{T} \ni z \mapsto A(z) \in M_2(\mathbb{C})$ such that $\det A(z) \equiv 1$ form a subgroup of $C(\mathbb{T}, \mathrm{GL}_2(\mathbb{C}))$ which we denote $\mathcal{SL}_2$.

For every $A(z)$ in $\mathcal{SL}_2$ there are $m \in \mathbb{Z}_+$, $K \in \mathbb{C}\backslash\{0\}$, and scalar-valued Fourier polynomials $u_1(z), \dots, u_m(z), l_1(z), \dots, l_m(z)$ such that

$$\begin{aligned} A(z) = \begin{pmatrix} K & 0 \\ 0 & K^{-1} \end{pmatrix} \cdot \begin{pmatrix} 1 & 0 \\ l_1(z) & 1 \end{pmatrix} \cdot \begin{pmatrix} 1 & u_1(z) \\ 0 & 1 \end{pmatrix} \\ \cdot \begin{pmatrix} 1 & 0 \\ l_2(z) & 1 \end{pmatrix} \cdot \begin{pmatrix} 1 & u_2(z) \\ 0 & 1 \end{pmatrix} \cdots \begin{pmatrix} 1 & 0 \\ l_m(z) & 1 \end{pmatrix} \cdot \begin{pmatrix} 1 & u_m(z) \\ 0 & 1 \end{pmatrix}. \end{aligned} \tag{100}$$

See [18]. This is the first step in the Daubechies–Sweldens lifting algorithm for the discrete wavelet transform. Thus the case $\det(A(z)) = 1$ gives a constructive lifting algorithm for wavelets, and such an algorithm has not been established in the $C(\mathbb{T}, \mathrm{GL}_2(\mathbb{C}))$ case. The decomposition could also be compared with Proposition 3.3 of [8], which was mentioned in connection with the proof of (24).

Recall the correspondence between matrix functions and wavelet filters: If $A\colon \mathbb{T} \to \mathrm{GL}_2(\mathbb{C})$ is a matrix function, then the corresponding dyadic wavelet filters are

$$m_i^{(A)}(z) = \sum_{j=0}^{1} A_{i,j}(z^2) z^j, \qquad i = 0, 1.$$

It follows that the two matrix functions A and B satisfy

$$A = \begin{pmatrix} 1 & 0 \\ l & 1 \end{pmatrix} B$$

for some l in the ring $\mathcal{F}$ of Fourier polynomials if and only if $m_0^{(A)} = m_0^{(B)}$ and $m_1^{(A)}(z) = m_1^{(B)}(z) + l(z^2) m_0^{(A)}(z)$.

Similarly note that the two matrix functions A and B satisfy

$$A = \begin{pmatrix} 1 & u \\ 0 & 1 \end{pmatrix} B$$

for some $u \in \mathcal{F}$ if and only if $m_1^{(A)} = m_1^{(B)}$ and $m_0^{(A)}(z) = m_0^{(B)}(z) + u(z^2)\, m_1^{(A)}(z)$.

Remark. The conclusion is that the wavelet algorithm for a general wavelet filter corresponding to a matrix function, say A, may be broken down in a sequence of zig-zag steps acting alternately on the high-pass and the low-pass signal components.

2.3. *Factorization theorems for matrix functions*

We mentioned that for matrix functions corresponding to finite impulse response (FIR) filters which are unitary, we need only the constant matrix (which is chosen such as to achieve the high-pass and low-pass conditions) and factors of the form

$$U_P(z) = zP + P^{\perp} \cong \left(\begin{array}{c|c} z & 0 \\ \hline 0 & 1 \end{array} \right)$$

where P is a rank-one projection in $\mathbb{C}^N$ and N is the scaling number of the subdivision.

Unfortunately, no such factorization theorem is available for the non-unitary FIR filters. But the matrix functions take values in the non-singular complex $N \times N$ matrices. The Sweldens–Daubechies factorization and the lifting algorithm serve as a substitute. There are still the general non-unimodular FIR-matrix functions where factorizations are so far a bit of a mystery. The matrix functions are called *polyphase matrices* in the engineering literature. The following summary serves as a classification theorem for the orthogonal wavelets of compact support: the wavelets correspond to FIR polyphase matrices which are unitary.

In summary, an algorithm to construct all the wavelet functions ψ of scale 2 with support in $[0, 2k+1]$ can be established as follows:

⟦1⟧ Pick k one-dimensional orthogonal projections $Q_1, \dots, Q_k$ in $M_2(\mathbb{C})$ and define the unitary-valued matrix function $A(z)$ on $\mathbb{T}$ by

$$A(z) = V(1 - Q_1 + zQ_1)(1 - Q_2 + zQ_2) \cdots (1 - Q_k + zQ_k), \quad (101)$$

where

$$V = \frac{1}{\sqrt{2}} \begin{pmatrix} 1 & 1 \\ 1 & -1 \end{pmatrix}. \quad (102)$$

Then each Q_j has the form

$$Q_j = \begin{pmatrix} \lambda_j & \sqrt{\lambda_j (1-\lambda_j)} e^{i\theta_j} \\ \sqrt{\lambda_j (1-\lambda_j)} e^{-i\theta_j} & 1-\lambda_j \end{pmatrix}, \tag{103}$$

where $\lambda_j \in [0,1]$ and $\theta_j \in [0, 2\pi)$. (See Proposition 3.3 of [8].)

⟦2⟧ Define the filters $m_0(z)$ and $m_1(z)$ by

$$m_i(z) = \sum_{j=0}^{N-1} z^j A_{ij}\left(z^N\right), \qquad i,j = 0,\ldots,N-1, \tag{104}$$

with $N = 2$.

⟦3⟧ Define $\hat{\varphi}$ by

$$\hat{\varphi}(t) = \prod_{k=1}^{\infty} \left(\frac{m_0\left(tN^{-k}\right)}{\sqrt{N}} \right). \tag{105}$$

If the condition

$$\mathrm{PER}\left(|\hat{\varphi}|^2\right)(t) := \sum_{n\in\mathbb{Z}} |\hat{\varphi}(t+2\pi n)|^2 = 1 \tag{106}$$

fails, then the algorithm stops.

⟦4⟧ If the condition (106) holds, one may alternatively define φ by the cascade algorithm

$$\varphi(x) = \sqrt{N} \sum_{n\in\mathbb{Z}} a_n \varphi(Nx-n), \tag{107}$$

$$\chi(x) = \begin{cases} 1, & 0 \le x < 1, \\ 0, & x \in \mathbb{R} \setminus [0,1\rangle, \end{cases} \tag{108}$$

$$M_a \colon \psi \longmapsto \sqrt{N} \sum_n a_n \psi(Nx-n). \tag{109}$$

⟦5⟧ The wavelet function ψ is then defined by

$$\psi_i(x) = \sqrt{N} \sum_{n\in\mathbb{Z}} a_n^{(i)} \varphi(Nx-n), \tag{110}$$

where $a_n^{(i)}$ are the Fourier coefficients of m_i,

$$m_i(z) = \sum_n a_n^{(i)} z^n, \tag{111}$$

and $z = e^{-it}$; this is the most general wavelet function with support in $[0, 2k+1]$.

⟦6⟧ All other wavelet functions with compact support can be obtained from the ones in ⟦5⟧ by integer translation.

2.3.1. *The case of polynomial functions [the polyphase matrix, joint work with Ola Bratteli]*

One problem occurring in the biorthogonal context which does not have an analogue in the orthogonal setting stems from the fact that the duality relations

$$\sum_{w^N=z} \overline{m_i(w)}\,\tilde{m}_j(w) = N\delta_{i,j} \qquad \text{for } i,j=0,\dots,N-1 \tag{112}$$

do not give any absolute restrictions on the size of m_i and $\tilde{m}_j$, e.g., a bound on the inner product of two vectors in $\mathbb{C}^N$ does not give a bound on the size of the vectors if they are not equal. This is reflected in the bi-Cuntz relations defined by m_i, $\tilde{m}_i$. Let us now define

$$(S_i f)(z) = m_i(z)\, f\left(z^N\right), \qquad (\tilde{S}_i f)(z) = \tilde{m}_i(z)\, f\left(z^N\right) \tag{113}$$

for $z \in \mathbb{T}$, $f \in L^2(\mathbb{T})$. Instead of the usual Cuntz relations, the S_i, $\tilde{S}_i$ now satisfy

$$S_i^* \tilde{S}_j = \delta_{i,j} 1, \tag{114}$$

$$\sum_i S_i \tilde{S}_i^* = 1. \tag{115}$$

If $A, \tilde{A} \in C(\mathbb{T}, \mathrm{GL}_N(\mathbb{C}))$ are the matrix-valued functions associated to m_i and $\tilde{m}_i$ by

$$m(z) = A\left(z^N\right) v(z), \qquad \tilde{m}(z) = \tilde{A}\left(z^N\right) v(z), \tag{116}$$

we compute

$$S_i^* S_j = (AA^*)_{j,i} \tag{117}$$

in the sense that $S_i^* S_j$ is contained in the commutative algebra of multiplication operators on $L^2(\mathbb{T})$ defined by $C(\mathbb{T})$, and $(AA^*)_{j,i} \in C(\mathbb{T})$. Correspondingly,

$$\tilde{S}_i^* \tilde{S}_j = (\tilde{A}\tilde{A}^*)_{j,i} \tag{118}$$

so all the operators $S_i^* S_j$, $\tilde{S}_i^* \tilde{S}_j$ are contained in the abelian algebra $C(\mathbb{T})$. We may introduce operators S, $\tilde{S}$ from

$$L^2(\mathbb{T})^N = \underset{0}{L^2(\mathbb{T})} \oplus \cdots \oplus \underset{N-1}{L^2(\mathbb{T})} \tag{119}$$

into $L^2(\mathbb{T})$ by

$$S = (S_0, S_1, \dots, S_{N-1}), \qquad \tilde{S} = (\tilde{S}_0, \dots, \tilde{S}_{N-1}) \tag{120}$$

and then S^* maps $L^2(\mathbb{T})$ into (119), etc., and the relations (114)–(118) take the form

$$\begin{cases} S^*\tilde{S} = 1, \text{ where 1 is the identity in } M_N(\mathbb{C}) \otimes C(\mathbb{T}), \\ S\tilde{S}^* = 1, \text{ where 1 is the identity in } C(\mathbb{T}), \end{cases} \tag{121}$$

$$\begin{cases} S^*S = AA^*, \\ \tilde{S}^*\tilde{S} = \tilde{A}\tilde{A}^*. \end{cases} \tag{122}$$

These relations say that all combinations of products of S and S^* with $\tilde{S}$ and $\tilde{S}^*$ lie in the algebra $M_N(\mathbb{C}) \otimes C(\mathbb{T})$. But in addition A and $\tilde{A}$ are matrix-valued functions on $\mathbb{T}$, so

$$AA^*\tilde{A}\tilde{A}^* = A\tilde{A}^* = 1 = \tilde{A}\tilde{A}^*AA^* \tag{123}$$

and hence

$$S^*S = \left(\tilde{S}^*\tilde{S}\right)^{-1} \tag{124}$$

and all the matrix-valued functions commute.

This discussion can be summarized by saying that the bi-Cuntz relations are much less rigid than the original Cuntz relations, i.e.:

Scholium 4: Given *any* bijective operator S from $L^2(\mathbb{T})^N$ into $L^2(\mathbb{T})$ one may define $\tilde{S} = (S^*)^{-1}$ and the bi-Cuntz relations (121) are satisfied. If, more specifically, S is given by (120) and (113), then operators $\tilde{S}_0, \dots, \tilde{S}_{N-1}$ exist such that the bi-Cuntz relations (114)–(115) are satisfied if and only if the operator $A \in M_N(\mathbb{C}) \otimes C(\mathbb{T})$ defined by (116) is invertible, in which case one must use $\tilde{A} = (A^*)^{-1}$, (116), and (113) to define $\tilde{S}_0, \dots, \tilde{S}_{N-1}$.

Let us now connect the filters to the wavelets. We have already defined the scaling functions φ, $\tilde{\varphi}$ and wavelet functions ψ_i, ψ_i, $i = 1, \dots, N$. The expansions for φ and $\tilde{\varphi}$ converge uniformly on compacts, thus $\hat{\varphi}$ and $\hat{\tilde{\varphi}}$ are continuous functions on $\mathbb{R}$. To decide that these functions are in $L^2(\mathbb{R})$ one again forms

$$f_\varphi(t) = \mathrm{PER}\left(|\hat{\varphi}|^2\right)(t) = \sum_{n\in\mathbb{Z}} |\hat{\varphi}(t - 2\pi n)|^2 \tag{125}$$

and $f_{\tilde{\varphi}}$ similarly, and one deduces again from the nonlinear intertwining relation

$$R^k(p(\psi_1, \psi_2)) = p\left(M_{m_0}^k \psi_1, M_{m_0}^k \psi_2\right), \qquad k \in \mathbb{N} \tag{126}$$

that

$$R_{m_0}(f_\varphi) = f_\varphi, \qquad R_{\tilde{m}_0}(f_{\tilde{\varphi}}) = f_{\tilde{\varphi}}. \tag{127}$$

2.3.2. *General results in mathematics on matrix functions*

In the standard case of the good old orthogonal wavelets in $L^2(\mathbb{R})$ of N subbands, we will look for functions $\psi_1, \dots, \psi_{N-1}$ in $L^2(\mathbb{R})$ such that, if k and n run independently over all the integers $\mathbb{Z}$, i.e., $-\infty < k, n < \infty$, then the countably infinite system of functions

$$\left\{ N^{k/2} \psi_i \left(N^k x - n \right) \,\middle|\, i = 1, \dots, N-1,\ k, n \in \mathbb{Z} \right\} \tag{128}$$

is an *orthonormal basis* in the Hilbert space $L^2(\mathbb{R})$. The second half of the word "orthonormal" refers to the restricting requirement that all the functions $\psi_1, \dots, \psi_{N-1}$ satisfy

$$\int_{\mathbb{R}} \left| \psi_i (x) \right|^2 dx = 1, \tag{129}$$

or stated more briefly,

$$\left\| \psi_i \right\|_{L^2(\mathbb{R})} = 1; \tag{130}$$

or yet more briefly,

$$\left\| \psi_i \right\| = 1. \tag{131}$$

From familiar properties of the Lebesgue measure on $\mathbb{R}$, it then follows that all the functions

$$\psi_{i,k,n} (x) := N^{k/2} \psi_i \left(N^k x - n \right), \qquad 1 \le i < N,\ k, n \in \mathbb{Z}, \tag{132}$$

satisfy the normalization, i.e., that

$$\left\| \psi_{i,k,n} \right\| = 1 \qquad \text{for all } i, k, n. \tag{133}$$

The functions (132) are said to be *orthogonal* if

$$\int_{\mathbb{R}} \overline{\psi_{i,k,n} (x)}\, \psi_{i',k',n'} (x)\, dx = 0 \tag{134}$$

whenever $(i, k, n) \neq (i', k', n')$. We say that the two triple indices are different if $i \neq i'$ or $k \neq k'$ or $n \neq n'$. If, for example, $i = i'$ and $k = k'$, then when the same function is translated by different amounts n and n', the two resulting functions are required to be orthogonal. It is an elementary geometric fact from the theory of Hilbert space that if the functions in (132) form an orthonormal basis, then for every function $f \in L^2(\mathbb{R})$, i.e., every measurable function f on $\mathbb{R}$ such that

$$\left\| f \right\|^2 = \int_{\mathbb{R}} \left| f (x) \right|^2 dx < \infty, \tag{135}$$

we have the identity

$$\|f\|^2 = \sum_{i,k,n} \left| \int_{\mathbb{R}} \overline{\psi_{i,k,n}(x)}\, f(x)\, dx \right|^2, \tag{136}$$

where the triple summation in (136) is over all configurations $1 \leq i < N$, $k, n \in \mathbb{Z}$. It is convenient to rewrite (136) in the following more compact form:

$$\|f\|^2 = \sum_{i,k,n} |\langle\, \psi_{i,k,n} \mid f \,\rangle|^2 . \tag{137}$$

Surprisingly, it turns out that (137) may hold even if the functions $\psi_{i,k,n}$ of (132) do not form an orthonormal basis. It may happen that one of the initial functions $\psi_1, \dots,$ or ψ_{N-1} satisfies $\|\psi_i\| < 1$, and yet that (137) holds for all $f \in L^2(\mathbb{R})$. These more general systems are still called wavelets, but since they are special, they are referred to as *tight frames*, as opposed to orthonormal bases. In either case, we will talk about a *wavelet expansion* of the form

$$f(x) = \sum_{i,k,n} \langle\, \psi_{i,k,n} \mid f \,\rangle\, \psi_{i,k,n}(x) . \tag{138}$$

It follows that the sum on the right-hand side in (138) converges in the norm of $L^2(\mathbb{R})$ for all functions f in $L^2(\mathbb{R})$ if (137) holds.

But there is a yet more general form of wavelets, called *biorthogonal.* The conditions on the functions $\psi_1, \dots, \psi_{N-1}$ are then much less restrictive than the orthogonality axioms. Hence these wavelets are more flexible and adapt better to a variety of applications, for example, to data compression, or to computer graphics. But the biorthogonality conditions are also a little more technical to state. We say that some given functions ψ_i, $i = 1, \dots, N-1$, in $L^2(\mathbb{R})$ are part of a biorthogonal wavelet system if there is a second system of functions $\tilde{\psi}_i$, $i = 1, \dots, N-1$, in $L^2(\mathbb{R})$, such that every $f \in L^2(\mathbb{R})$ admits a representation

$$f(x) = \sum_{i,k,n} \langle\, \psi_{i,k,n} \mid f \,\rangle\, \tilde{\psi}_{i,k,n}(x) = \sum_{i,k,n} \langle\, \tilde{\psi}_{i,k,n} \mid f \,\rangle\, \psi_{i,k,n}(x) , \tag{139}$$

and

$$\tilde{\psi}_{i,k,n}(x) = N^{k/2} \tilde{\psi}_i \left(N^k x - n\right) . \tag{140}$$

In the standard normalized case where $\langle\, \psi_i \mid \tilde{\psi}_i \,\rangle = 1$, then you will notice that condition (137) turns into

$$\|f\|^2 = \sum_{i,k,n} \overline{\langle\, \psi_{i,k,n} \mid f \,\rangle}\, \langle\, \tilde{\psi}_{i,k,n} \mid f \,\rangle \tag{141}$$

for all $f \in L^2(\mathbb{R})$.

The orthogonal wavelets correspond to matrix functions $\mathbb{T} \to \mathrm{U}_N(\mathbb{C})$, while the wider class of biorthogonal wavelets corresponds to the much bigger group of matrix functions $\mathbb{T} \to \mathrm{GL}_N(\mathbb{C})$, via the associated wavelet filters. You may ask, why bother with the more technical-looking biorthogonal systems? It turns out that they are forced on us by the engineers. They tell us that the real world is not nearly as orthogonal as the mathematicians would like to make it out to be. There is a paucity of symmetric orthogonal wavelets, and symmetry ("linear phase") is prized by engineers and workers in image processing, where the more general wavelet families and their duality play a crucial role. Now what if we could change the biorthogonal wavelets into the orthogonal ones, and still keep the essential spectral properties intact? Then everyone will be happy. This last chapter shows that it is possible, and even in a fairly algorithmic fashion, one that is amenable to computations.

Wavelet filters may be understood as matrix functions, i.e., functions from the one-torus $\mathbb{T} \subset \mathbb{C}$ into some group of invertible matrices. If the scale number is N, then there are three such matrix groups which are especially relevant for wavelet analysis:

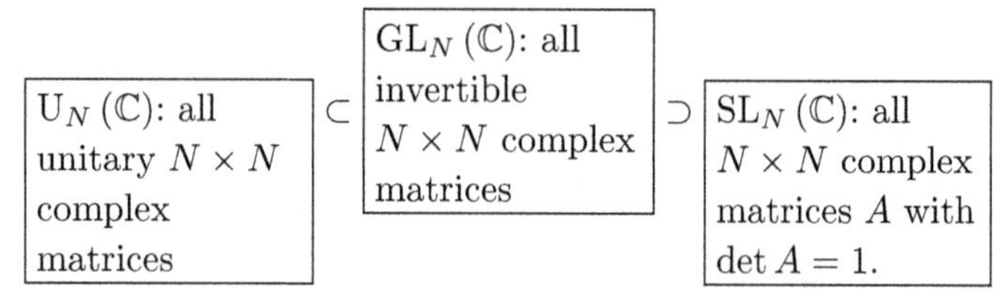

It is possible to reduce some questions in the GL_N case to better understood results for $\mathrm{U}_N(\mathbb{C})$; see Chapter 6 of [9]. The SL_2 case is especially interesting in view of Daubechies–Sweldens lifting for dyadic wavelets; see § 2.2.4.

2.3.3. *Connection between matrix functions and wavelets*

Definitions: A function, or a distribution, φ satisfying (107) is said to be *refinable*, the equation (107) is called the *refinement equation*, or also, as noted above, the "scaling identity", and φ is called the scaling function. The coefficients a_n of (107) are called the *masking coefficients.*

We will mainly concentrate on the case when the set $\{a_n\}$ is finite. But in general, a function $\varphi \in L^2(\mathbb{R})$ is said to be refinable with scale number N if $\varphi(x/N)$ is in the L^2-closed linear span of the translates $\{\varphi(x-k)\}_{k\in\mathbb{Z}} \subset L^2(\mathbb{R})$; see, e.g., [32, 81, 82, 83].

Since there are refinement operations which are more general than scaling (see for example [23]), there are variations of (107) which are correspondingly more general, with regard to both the refinement steps that are used and the dimension of the spaces. The term "scaling identity" is usually, but not always, reserved for (107), while more general refinements lead to "refinement equations". However, (107) often goes under both names. The vector versions of the identities get the prefix "multi-", for example *multi-scaling* and *multiwavelet.*

If m_0 satisfies a condition for obtaining orthogonal wavelets,

$$\sum_{w^N=z} \left| m_0 \left(w \right) \right|^2 = N, \tag{142}$$

together with the normalization

$$m_0 \left(1 \right) = \sqrt{N}, \tag{143}$$

then (107) has a solution φ in $L^2 \left(\mathbb{R} \right)$ which can be obtained by taking the inverse Fourier transform of the product expansion

$$\hat{\varphi} \left(t \right) = \prod_{k=1}^{\infty} \left(\frac{m_0 \left(t N^{-k} \right)}{\sqrt{N}} \right). \tag{144}$$

(Here and later we use the convention that if $m \left(z \right)$ is a function of $z \in \mathbb{T}$, then $m \left(t \right) = m \left(e^{-it} \right)$.) That (144) gives a solution φ of (107) follows from the relation

$$\hat{\varphi} \left(t \right) = \frac{1}{\sqrt{N}} m_0 \left(\frac{t}{N} \right) \hat{\varphi} \left(\frac{t}{N} \right). \tag{145}$$

2.3.3.1. *Multiresolution wavelets*

We mentioned that there is a direct connection between $m_0 = \sum a_n z^n$ and the scaling function φ on $\mathbb{R}$ given in (34), (107), and (144). There is a similar correspondence between the high-pass filters m_i and the wavelet generators $\psi_i \in L^2 \left(\mathbb{R} \right)$. In the *biorthogonal* case, there is a second system $\tilde{m}_i \leftrightarrow \tilde{\psi}_i$ and the two systems

$$\left\{ N^{\frac{j}{2}} \psi_i \left(N^j x - k \right) \right\} \quad \text{and} \quad \left\{ N^{\frac{j'}{2}} \tilde{\psi}_{i'} \left(N^{j'} x - k' \right) \right\},$$
$$i, i' \in \left\{ 1, 2, \dots, N-1 \right\},\ j, j', k, k' \in \mathbb{Z}, \tag{146}$$

then form a dual wavelet basis, or dual wavelet frame for $L^2 \left(\mathbb{R} \right)$ in the sense of [15], Chapter 5. We considered this biorthogonal case in more detail in § 2.3.1 above. Much more detail can be found in Chapter 6 of [9].

The idea of constructing maximally smooth wavelets when some side conditions are specified has been central to much of the activity in wavelet analysis and its applications since the mid-1980's. As a supplement to [15], the survey article [79] is enjoyable reading. The paper [58] treats the issue in a more specialized setting and is focussed on the moment method. Some of the early applications to data compression and image coding are done very nicely in [34], [84], and [33]. An interesting, related but different, algebraic and geometric approach to the problem is offered in [67].

We now turn to an interesting variation of this setup, which includes higher dimensions, i.e., when the Hilbert space is $L^2(\mathbb{R}^d)$, $d = 2, 3, \ldots$. Staying for the moment with $d = 1$, and N fixed, we will take the viewpoint of what is called *resolutions*, but here understood in a broad sense of closed subspaces: A closed linear subspace $\mathcal{V} \subset L^2(\mathbb{R})$ is said to be an N-resolution if it is invariant under the unitary operator

$$U = U_N \colon f \longmapsto N^{-\frac{1}{2}} f\left(\frac{x}{N}\right), \tag{147}$$

i.e., if U maps $\mathcal{V}$ into a proper subspace of itself. The subspace $\mathcal{V}$ is said to be *translation invariant* if

$$f \in \mathcal{V} \iff f(\cdot - k) \in \mathcal{V} \qquad \text{for all } k \in \mathbb{Z}. \tag{148}$$

If there is a function φ such that $\mathcal{V} = \mathcal{V}_\varphi$ is the closed linear span of

$$\{\varphi(\cdot - k) \mid k \in \mathbb{Z}\}, \tag{149}$$

then clearly $\mathcal{V}$ is translation invariant. The translation-invariant resolution subspaces $\mathcal{V}$ are actively studied and reasonably well understood. If $\mathcal{V}$ is of the form $\mathcal{V}_\varphi$ in (149), then we say that it is *singly generated*, and that φ is a scaling function of scale N.

2.3.3.2. *Generalized multiresolutions [joint work with L. Baggett, K. Merrill, and J. Packer]*

The case when the resolution subspace $\mathcal{V}$ is not singly generated is also interesting, and these resolution subspaces are frequently called *generalized multiresolution subspaces* (GMRA). There is much current and very active research on them; see, for example, [4], [60], [5], [30], [31], [81], and [45]. The case when $\mathcal{V}$ is not singly generated as a resolution subspace of scale $N > 2$, i.e., when $\mathcal{V}$ is not of the form (149), occurs in the study of *wavelet sets*. A wavelet set in $\mathbb{R}^d$ is defined relative to an expansive $d \times d$ matrix $\mathbf{N}$ over $\mathbb{Z}$. A subset $E \subset \mathbb{R}^d$ is said to be an $\mathbf{N}$-wavelet set if there is a single wavelet function $\psi \in L^2(\mathbb{R}^d)$ such that $\hat{\psi} = \chi_E$. Specifically, the condition states that the family

$$\left\{ |\det \mathbf{N}|^{j/2} \psi\left(\mathbf{N}^j x - \mathbf{k}\right) : j \in \mathbb{Z},\ \mathbf{k} \in \mathbb{Z}^d \right\} \tag{150}$$

is an orthonormal basis for $L^2(\mathbb{R}^d)$. This can be checked to be equivalent to the combined set of two tiling properties for E as a subset of $\mathbb{R}^d$:

(a) the family of subsets $\left\{ \mathbf{N}^j E : j \in \mathbb{Z} \right\}$ tiles $\mathbb{R}^d$;
(b) the translates $\left\{ E + 2\pi\mathbf{k} : \mathbf{k} \in \mathbb{Z}^d \right\}$ tile $\mathbb{R}^d$.

We define tiling by the requirement that the sets in the family have overlap at most of measure zero relative to Lebesgue measure on $\mathbb{R}^d$. Similarly, the union

$$\mathbb{R}^d = \bigcup_{j \in \mathbb{Z}} \mathbf{N}^j E = \bigcup_{\mathbf{k} \in \mathbb{Z}^d} E + 2\pi\mathbf{k} \tag{151}$$

is understood to be only up to measure zero.

It is easy to see that compactly supported wavelets in $L^2(\mathbb{R}^d)$ are MRA wavelets, while most wavelets $\psi = (\chi_E)^{\vee}$ from wavelet sets E are not. These wavelets are typically (but not always) frequency localized.

The main difference between the GMRA (stands for generalized multiresolution analysis) wavelets and the more traditional MRA ones may be understood in terms of multiplicity. Both come from a fixed resolution subspace $\mathcal{V}_0 \subset L^2(\mathbb{R}^d)$ which is invariant under the translations $\left\{ T_n : n \in \mathbb{Z}^d \right\}$ where

$$(T_n f)(x) := f(x - n) \qquad \text{for } x \in \mathbb{R}^d \text{ and } n \in \mathbb{Z}^d. \tag{152}$$

Hence $\{T_n|_{\mathcal{V}_0}\}_{n \in \mathbb{Z}^d}$ is a unitary representation of $\mathbb{Z}^d$ on the Hilbert space $\mathcal{V}_0$. As a result of Stone's theorem, we find that there are subsets

$$E_1 \supset E_2 \supset \cdots \supset E_j \supset \cdots$$

of $\mathbb{T}^d$ such that the spectral measure of the (restricted) representation has multiplicity $\geq j$ on the subset E_j, $j = 1, 2, \ldots$. It can be checked that the projection-valued spectral measure is absolutely continuous. Moreover, there is an intertwining unitary operator

$$J \colon \mathcal{V}_0 \longrightarrow \sum_{j \geq 1}^{\oplus} L^2(E_j) \tag{153}$$

such that

$$P_{L^2(E_j)} J T_n f(z) = z^n (Jf)(z) \tag{154}$$

holds for all $f \in \mathcal{V}_0$ and $z \in E_j$. We may then consider the functions $\varphi_j \in \mathcal{V}_0$ $(\subset L^2(\mathbb{R}^d))$ defined by

$$\varphi_j := J^{-1}(0, \dots, 0, \underbrace{\chi_{E_j}}_{j\text{'th place}}, 0, 0, \dots). \tag{155}$$

It was proved by Baggett and Merrill [5] that $\{\varphi_j : j \geq 1\}$ generates a normalized tight frame for $\mathcal{V}_0$: specifically, that

$$\sum_{j \geq 1} \sum_{n \in \mathbb{Z}^d} |\langle T_n \varphi_j \mid f \rangle|^2 = \|f\|^2_{L^2(\mathbb{R}^d)} \tag{156}$$

holds for all $f \in \mathcal{V}_0$.

Treating $(\varphi_1, \varphi_2, \dots)$ as a vector-valued function, denoted simply by φ, we see that there is a matrix function

$$H : \mathbb{T}^d \longrightarrow \text{(complex square matrices)}$$

such that

$$\hat{\varphi}\left(\mathbf{N}^{\text{tr}} t\right) = H\left(e^{it}\right) \hat{\varphi}(t), \tag{157}$$

(the superscript tr is annotated "for transpose"), where $t = (t_1, \dots, t_d) \in \mathbb{R}^d$, and $e^{it} := \left(e^{it_1}, e^{it_2}, \dots, e^{it_d}\right)$.

But this method takes the Hilbert space $L^2(\mathbb{R}^d)$ as its starting point, and then proceeds to the construction of wavelet filters in the form (157). Our current joint work with Baggett, Merrill, and Packer reverses this. It begins with a matrix function H defined on $\mathbb{T}^d$, and then offers subband conditions on the matrix function which allow the construction of a GMRA for $L^2(\mathbb{R}^d)$ with generator $\varphi = (\varphi_1, \varphi_2, \dots)$ given by (157). So the Hilbert space $L^2(\mathbb{R}^d)$ shows up only at the end of the construction, in the conclusions of the theorems.

2.3.4. *Matrix completion*

In using the polyphase matrices, one may only have the first few rows, and be faced with the problem of completing to get the entire function A from a torus into the matrices of the desired size. The case when only the first row is given, say corresponding to a specified low-pass filter, is treated in [9] and [8], and we refer the reader to the references given there, especially [42], [69], [70], and [87].

The wavelet transfer operator is used in a variety of wavelet applications not covered here, or only touched upon tangentially: stability of refinable functions, regularity, approximation order, unitary matrix extension principles, to mention only a few. The reader is referred to the following papers for more details on these subjects: [17], [75], [76], [41], [74], [71], [43], [78], [73], [59], [72], [2], and [3].

The unitary extension principle (UEP) was first published in the paper [72] by A. Ron and Z. Shen. The result is due jointly to the two Ron and Shen. In fact, they not only formulated and proved the UEP for the refinable PSI case, but also in the refinable FSI case. The main added contribution in [17] is the related but different oblique extension principle (OEP). The oblique extension principle (OEP) in [17] is a generalization of the UEP. It yields powerful results on the approximation power of the truncated tight frame series.

The unitary extension principle (UEP) of [72] (see also [17]) involves the interplay between a finite set of filters (functions on $\mathbb{R}/\mathbb{Z}$), and a corresponding tight frame (alias Parseval frame) in $L^2(\mathbb{R}^d)$.

For the sake of illustration, let us take $d = 1$, and scaling number $N = 2$, i.e., the case of dyadic framelets. Naturally, the notion of tight frame is weaker than that of an orthonormal basis (ONB), and it is shown in [17] that when a system of wavelet filters m_i, $i = 0, 1, \ldots, r$ is given (m_0 must be low-pass), then the orthogonality condition on the m_i's always gets us a framelet in $L^2(\mathbb{R})$, i.e., the functions ψ_i corresponding to the high-pass filters m_i, $i = 1, \ldots, r$, generate a tight frame for $L^2(\mathbb{R})$, also called a framelet. The correspondence m_i to ψ_i is called the UEP in [72] (see also [17]).

The orthogonality condition for m_i, $i = 0, 1, \ldots, r$, referred to in the UEP is simply this: Form an $(r+1)$-by-2 matrix-valued function $F(x)$ by using $m_i(x)$, $i = 0, 1, \ldots, r$ in the first column, and the translates of the m_i's by a half period, i.e., $m_i(x + 1/2)$, $i = 0, 1, \ldots, r$ in the second. The condition on this matrix function $F(x)$ is that the two columns are orthogonal and have unit norm in ℓ^2 for all x. Note that we still get the unitary matrix functions acting on these systems, in the way we outlined above. But there is redundancy as the unitary matrices are $(r+1)$-by-$(r+1)$. The reader is referred to [17] for further details.

We emphasize that several of these, and other related topics, invite the kind of probabilistic tools that we have stressed here. But a more systematic discussion is outside the scope of this brief set of notes. We only hope to offer a modest introduction to a variety of more specialized topics.

Remark 5: The orthogonality condition for m_i, $i = 0, 1, \ldots, r$, may be stated in terms of the operators S_i from equation (73), $N = 2$. For each $i = 0, 1, \ldots, r$, define an operator on $L^2(\mathbb{R}/\mathbb{Z})$ as in (73). Then the arguments from Section 2 show that the orthogonality condition for m_i, $i = 0, 1, \ldots, r$, i.e., the UEP condition, is equivalent to the operator identity (72) where the summation now runs from 0 to r. Operator systems S_i satisfying (72) are called row-isometries.

Remark 6: There are two properties of the low-pass filter m_0 which we have glossed over. First, m_0 must be such that the corresponding scaling function φ is in $L^2(\mathbb{R})$. Without an added condition on m_0, φ might only be a distribution. Secondly, when the dyadic scaling in $L^2(\mathbb{R})$ is restricted to the resolution subspace $V_0(\varphi)$, the corresponding unitary part must be zero. These two issues are addressed in [2], [3], and [17].

2.3.5. *Connections between matrix functions and signal processing*

Since our joint work with Baggett, Merrill, and Packer on the GMRA wavelets is still in progress, we restrict the discussion of matrix functions here to the MRA case.

The two groups of matrix functions $C(\mathbb{T}, \mathrm{U}_N(\mathbb{C}))$ and $C(\mathbb{T}, \mathrm{GL}_N(\mathbb{C}))$, i.e., the continuous functions from the torus into the respective groups, enter wavelet analysis via the associated wavelet filters $(m_i)_{i=0}^{N-1}$.

In [9] (see also § 1.1.3 above), we give the details of the multiple correspondence between:

(i) matrix functions, $A\colon \mathbb{T} \to \mathrm{GL}_N(\mathbb{C})$,
(ii) high- and low-pass wavelet filters m_i, $\tilde{m}_{i'}$, $i, i' = 0, 1, \ldots, N-1$, and
(iii) wavelet generators ψ_i, $\tilde{\psi}_{i'}$, $i, i' = 1, \ldots, N-1$, together with scaling functions φ, $\tilde{\varphi}$.

In particular,

$$A_{i,j}(z) = \frac{1}{N} \sum_{w^N = z} m_i(w)\, w^{-j}, \qquad z \in \mathbb{T}, \tag{158}$$

$$\left(A^{-1}\right)_{i,j} = \frac{1}{N} \sum_{w^N = z} \overline{\tilde{m}_j(w)}\, w^{i}, \qquad z \in \mathbb{T}. \tag{159}$$

The dependence of the $L^2(\mathbb{R})$-functions in (iii) on the group elements A from (i) gives rise to homotopy properties. The standard orthogonal wavelets represent the special case when $m_i = \tilde{m}_i$, or equivalently, $A(z) = \left((A(z))^*\right)^{-1}$, $z \in \mathbb{T}$. Hence, the matrix functions are unitary in this case.

The scaling/wavelet functions $\varphi, \psi_1, \dots, \psi_{N-1}$ with support on a fixed compact interval, say $[0, kN+1]$, $k = 0, 1, \dots$, can be parameterized with a finite number of parameters since the unitary-valued function $z \to A(z)$ in (158) then is a polynomial in z of degree at most $k(N-1)$. It is well-known folklore from computer-generated pictures that the shape of the scaling/wavelet functions depends continuously on these parameters; see Figures 1.1–1.7 in [9] and [86].

The scaling function $\varphi \in L^2(\mathbb{R})$ of (107) is illustrated there, in the case $N = 2$, and for orthogonal $\mathbb{Z}$-translates, i.e., the case (142). These pictures illustrate the dependence of φ on the masking coefficients (a_n) in the case of [86]:

$$
\begin{aligned}
a_0 &= (\eta_0 - \eta_1 - \eta_2 + \eta_3 + \eta_4)/4, \\
a_1 &= (\eta_0 + \eta_1 - \eta_2 + \eta_3 - \eta_4)/4, \\
a_2 &= (\eta_0 - \eta_3 - \eta_4)/2, \\
a_3 &= (\eta_0 - \eta_3 + \eta_4)/2, \\
a_4 &= (\eta_0 + \eta_1 + \eta_2 + \eta_3 + \eta_4)/4, \\
a_5 &= (\eta_0 - \eta_1 + \eta_2 + \eta_3 - \eta_4)/4,
\end{aligned}
\tag{160}
$$

where

$$
\begin{aligned}
&\eta_0 = 1/\sqrt{2}, \\
&\eta_1 = (\cos 2\theta + \cos 2\rho)/\sqrt{2}, \qquad \eta_2 = (\sin 2\theta + \sin 2\rho)/\sqrt{2}, \\
&\eta_3 = \cos(2\theta - 2\rho)/\sqrt{2}, \qquad \eta_4 = \sin(2\theta - 2\rho)/\sqrt{2}.
\end{aligned}
\tag{161}
$$

These formulas arise from an independent pair of rotations by angles θ and ρ of two "spin vectors", i.e., by taking the matrix function A in (158) unitary, $\mathbb{T} \ni z \to A_{\theta,\rho}(z) \in \mathrm{U}_2(\mathbb{C})$, and setting

$$
A(z) = V(Q_\theta^\perp + zQ_\theta)(Q_\rho^\perp + zQ_\rho) = VU_\theta(z)U_\rho(z)
\tag{162}
$$

with

$$
V = \frac{1}{\sqrt{2}} \begin{pmatrix} 1 & 1 \\ 1 & -1 \end{pmatrix},
\tag{163}
$$

$$
\begin{aligned}
Q_\theta &= \begin{pmatrix} \cos^2\theta & \cos\theta\sin\theta \\ \cos\theta\sin\theta & \sin^2\theta \end{pmatrix} \\
&= \frac{1}{2}\left(\begin{pmatrix} 1 & 0 \\ 0 & 1 \end{pmatrix} + \begin{pmatrix} \cos 2\theta & \sin 2\theta \\ \sin 2\theta & -\cos 2\theta \end{pmatrix} \right),
\end{aligned}
\tag{164}
$$

and the orthogonal complement to the one-dimensional projection Q_θ,

$$Q_\theta^\perp = Q_{\theta+(\pi/2)}. \tag{165}$$

With the coefficients $a_0, a_1, a_2, a_3, a_4, a_5$ given by (160), the algorithmic approach to graphing the solution φ to the scaling identity (107) is as follows (see [46], [86] for details): the relation (107) for $N = 2$ is interpreted as giving the values of the left-hand φ by an operation performed on those of the φ on the right, and a binary digit inversion transforms this into the form

$$\mathbf{f}'_{k+1}\left(x + \frac{1}{2^{k+1}}\right) = \mathbf{A}\mathbf{f}_k(x), \tag{166}$$

where $\mathbf{A}$ is the 2×3 matrix $\mathbf{A}_{i,j} = \sqrt{2}a_{4+i-2j}$ constructed from the coefficients in (107), and $\mathbf{f}_j$ and $\mathbf{f}'_j$ are the vector functions

$$\mathbf{f}_j(x) = \begin{pmatrix} \varphi\left(x - \frac{2}{2^j}\right) \\ \varphi\left(x - \frac{1}{2^j}\right) \\ \varphi(x) \end{pmatrix}, \qquad \mathbf{f}'_j(x) = \begin{pmatrix} \varphi\left(x - \frac{1}{2^j}\right) \\ \varphi(x) \end{pmatrix}. \tag{167}$$

The signal processing aspect can be understood from the description of subband filters in the analysis and synthesis of time signals, or more general signals for images. In either case, we have two subband systems $m = (m_0, m_1, \dots)$ and $\tilde{m} = (\tilde{m}_0, \tilde{m}_1, \dots)$ where the functions

$$m_j(z) = \sum_n a_n^{(j)} z^n \quad \text{and} \quad \tilde{m}_j(z) = \sum_n \tilde{a}_n^{(j)} z^n$$

are the generating functions defined from the filter coefficients $\left(a_n^{(j)}\right)$ and $\left(\tilde{a}_n^{(j)}\right)$, $n \in \mathbb{Z}^d$.

Acknowledgments

We are happy to thank the organizing committee at the National University of Singapore for all their dedicated work in planning and organizing a successful conference, of which this tutorial is a part. We especially thank Professors Wai Shing Tang, Judith Packer, Zuowei Shen, and the head of the Department of Mathematics of the NUS for all their work in making my visit to Singapore possible. We thank the Institute for Mathematical Sciences at the NUS, and the US National Science Foundation under grants DMS-9987777, DMS-0139473(FRG), for partial financial support in the preparation of these lecture notes. We discussed various parts of the mathematics with our colleagues, Professors Larry Baggett, David Larson, Ola Bratteli, Kathy Merrill, Judy Packer, and we thank them for their encouragements and suggestions. The typesetting and graphics were expertly done at the University of Iowa by Brian Treadway. We also thank Brian Treadway for a number of corrections, and for very helpful suggestions, leading to improvements of the presentation.

References

1. P. Auscher, Solution of two problems on wavelets, J. Geom. Anal. 5 (1995), 181–236.
2. L. W. Baggett, P. E. T. Jorgensen, K. D. Merrill, and J. A. Packer, Construction of Parseval wavelets from redundant filter systems, preprint, 2003, submitted to J. Amer. Math. Soc., arXiv:math.CA/0405301.
3. L. W. Baggett, P. E. T. Jorgensen, K. D. Merrill, and J. A. Packer, An analogue of Bratteli-Jorgensen loop group actions for GMRA's, in Wavelets, Frames, and Operator Theory (College Park, MD, 2003), ed. C. Heil, P. E. T. Jorgensen, and D. R. Larson, Contemp. Math., vol. 345, American Mathematical Society, Providence, 2004, pp. 11–25.
4. L. W. Baggett and D. R. Larson (eds.), The Functional and Harmonic Analysis of Wavelets and Frames: Proceedings of the AMS Special Session on the Functional and Harmonic Analysis of Wavelets Held in San Antonio, TX, January 13–14, 1999, Contemp. Math., vol. 247, American Mathematical Society, Providence, 1999.
5. L. W. Baggett and K. D. Merrill, Abstract harmonic analysis and wavelets in $\mathbf{R}^n$, in The Functional and Harmonic Analysis of Wavelets and Frames (San Antonio, 1999), ed. L.W. Baggett and D.R. Larson, Contemp. Math., vol. 247, American Mathematical Society, Providence, 1999, pp. 17–27.
6. J. J. Benedetto, Ten books on wavelets, SIAM Rev. 42 (2000), 127–138.
7. O. Bratteli and P. E. T. Jorgensen, Convergence of the cascade algorithm at irregular scaling functions, in The Functional and Harmonic Analysis of Wavelets and Frames (San Antonio, 1999), ed. L. W. Baggett and D. R. Larson, Contemp. Math., vol. 247, American Mathematical Society, Providence, 1999, pp. 93–130.
8. O. Bratteli and P. E. T. Jorgensen, Wavelet filters and infinite-dimensional unitary groups, in Wavelet Analysis and Applications (Guangzhou, China, 1999), ed. D. Deng, D. Huang, R.-Q. Jia, W. Lin, and J. Wang, AMS/IP Studies in Advanced Mathematics, vol. 25, American Mathematical Society, Providence, International Press, Boston, 2002, pp. 35–65.
9. O. Bratteli and P. E. T. Jorgensen, Wavelets through a Looking Glass: The World of the Spectrum, Applied and Numerical Harmonic Analysis, Birkhäuser, Boston, 2002.
10. B. Brenken and P. E. T. Jorgensen, A family of dilation crossed product algebras, J. Operator Theory 25 (1991), 299–308.
11. A. Cohen, I. Daubechies, and P. Vial, Wavelets on the interval and fast wavelet transforms, Appl. Comput. Harmon. Anal. 1 (1993), 54–81.

12. R. R. Coifman and M. V. Wickerhauser, Wavelets and adapted waveform analysis: A toolkit for signal processing and numerical analysis, in Different Perspectives on Wavelets (San Antonio, TX, 1993), ed. I. Daubechies, Proc. Sympos. Appl. Math., vol. 47, American Mathematical Society, Providence, 1993, pp. 119–153.
13. J. Cuntz, Simple C^*-algebras generated by isometries, Comm. Math. Phys. 57 (1977), 173–185.
14. X. Dai and D. R. Larson, Wandering vectors for unitary systems and orthogonal wavelets, Mem. Amer. Math. Soc. 134 (1998), no. 640.
15. I. Daubechies, Ten Lectures on Wavelets, CBMS-NSF Regional Conf. Ser. in Appl. Math., vol. 61, SIAM, Philadelphia, 1992.
16. I. Daubechies, A. Grossmann, and Y. Meyer, Painless nonorthogonal expansions, J. Math. Phys. 27 (1986), 1271–1283.
17. I. Daubechies, B. Han, A. Ron, and Z. Shen, Framelets: MRA-based constructions of wavelet frames, Appl. Comput. Harmon. Anal. 14 (2003), 1–46.
18. I. Daubechies and W. Sweldens, Factoring wavelet transforms into lifting steps, J. Fourier Anal. Appl. 4 (1998), 247–269.
19. D. E. Dutkay and P. E. T. Jorgensen, Wavelets on fractals, preprint, University of Iowa, 2003, to appear in Rev. Mat. Iberoamericana, arXiv:math.CA/0305443.
20. D. E. Dutkay and P. E. T. Jorgensen, Martingales, endomorphisms, and covariant systems of operators in Hilbert space, preprint, University of Iowa, 2004, submitted to Amer. J. Math., arXiv:math.CA/0407330.
21. D. E. Dutkay and P. E. T. Jorgensen, Operators, martingales, and measures on projective limit spaces, preprint, University of Iowa, 2004, submitted to Trans. Amer. Math. Soc., arXiv:math.CA/0407517.
22. D. E. Dutkay and P. E. T. Jorgensen, Disintegration of projective measures, preprint, University of Iowa, 2004, submitted to Proc. Amer. Math. Soc., arXiv:math.CA/0408151.
23. N. Dyn, D. Leviatan, D. Levin, and A. Pinkus (eds.), Multivariate Approximation and Applications, Cambridge University Press, Cambridge, England, 2001.
24. D. Esteban and C. Galand, Application of quadrature mirror filters to split band voice coding systems, in IEEE International Conference on Acoustics, Speech, and Signal Processing (Washington, DC, May 1977), Institute of Electrical and Electronics Engineers, Piscataway, NJ, 1977, pp. 191–195.
25. A. Fijany and C. P. Williams, Quantum wavelet transforms: Fast algorithms and complete circuits, in Quantum Computing and Quantum Communications (Palm Springs, CA, 1998), ed. C. P. Williams, Lecture Notes in Computer Science, vol. 1509, Springer, Berlin, 1999, pp. 10–33.
26. M. H. Freedman, Poly-locality in quantum computing, Found. Comput. Math. 2 (2002), 145–154.
27. B. Fuglede, Commuting self-adjoint partial differential operators and a group-theoretic problem, J. Funct. Anal. 16 (1974), 101–121.
28. G. Garrigós, Connectivity, homotopy degree, and other properties of α-localized wavelets on $\mathbb{R}$, Publ. Mat. 43 (1999), 303–340.

29. A. Haar, Zur Theorie der orthogonalen Funktionensysteme, Math. Ann. 69 (1910), 331–371.
30. D. Han, D. R. Larson, M. Papadakis, and Th. Stavropoulos, Multiresolution analyses of abstract Hilbert spaces and wandering subspaces, in The Functional and Harmonic Analysis of Wavelets and Frames (San Antonio, 1999), ed. L. W. Baggett and D. R. Larson, Contemp. Math., vol. 247, American Mathematical Society, Providence, 1999, pp. 259–284.
31. X. M. He, L. X. Shen, and Z. Shen, A data-adaptive knot selection scheme for fitting splines, IEEE Signal Processing Letters 8 (2001), 137–139.
32. C. Heil, G. Strang, and V. Strela, Approximation by translates of refinable functions, Numer. Math. 73 (1996), 75–94.
33. P. N. Heller, J. M. Shapiro, and R. O. Wells, Optimally smooth symmetric quadrature mirror filters for image coding, in Wavelet Applications II (Orlando, 1995), ed. H.H. Szu, Proceedings of SPIE, vol. 2491, Society of Photo-optical Instrumentation Engineers, Bellingham, WA, 1995, pp. 119–130.
34. P. N. Heller, V. Strela, G. Strang, P. Topiwala, C. Heil, and L. S. Hills, Multiwavelet filter banks for data compression, in IEEE International Symposium on Circuits and Systems, 1995 (ISCAS '95), Institute of Electrical and Electronics Engineers, New York, 1995, vol. 3, pp. 1796–1799.
35. E. Hernandez and G. Weiss, A First Course on Wavelets, Studies in Advanced Mathematics, CRC Press, Boca Raton, Florida, 1996.
36. B. B. Hubbard, The World According to Wavelets: The Story of a Mathematical Technique in the Making, second ed., A. K. Peters, Wellesley, MA, 1998.
37. J. E. Hutchinson, Fractals and self similarity, Indiana Univ. Math. J. 30 (1981), 713–747.
38. W.-L. Hwang and S. Mallat, Characterization of self-similar multifractals with wavelet maxima, Appl. Comput. Harmon. Anal. 1 (1994), 316–328.
39. A. Iosevich, N. Katz, and T. Tao, The Fuglede spectral conjecture holds for convex planar domains, Math. Res. Lett. 10 (2003), 559–569.
40. A. Iosevich and S. Pedersen, Spectral and tiling properties of the unit cube, Internat. Math. Res. Notices 1998 (1998), 819–828.
41. R.-Q. Jia, Q. Jiang, and Z. Shen, Convergence of cascade algorithms associated with nonhomogeneous refinement equations, Proc. Amer. Math. Soc. 129 (2001), 415–427.
42. R. Q. Jia and Z. Shen, Multiresolution and wavelets, Proc. Edinburgh Math. Soc. (2) 37 (1994), 271–300.
43. Q. Jiang and Z. Shen, On existence and weak stability of matrix refinable functions, Constr. Approx. 15 (1999), 337–353.
44. P. E. T. Jorgensen, Spectral theory of finite-volume domains in $\mathbb{R}^n$, Adv. in Math. 44 (1982), 105–120.
45. P. E. T. Jorgensen, Ruelle operators: Functions which are harmonic with respect to a transfer operator, Mem. Amer. Math. Soc. 152 (2001), no. 720.
46. P. E. T. Jorgensen, Minimality of the data in wavelet filters, Adv. Math. 159 (2001), 143–228.

47. P. E. T. Jorgensen, Invited featured book review of *An Introduction to Wavelet Analysis* by D. F. Walnut, Applied and Numerical Harmonic Analysis, Birkhäuser, 2002, Bull. Amer. Math. Soc. (N.S.) 40 (2003), 421–427.
48. P. E. T. Jorgensen, Matrix factorizations, algorithms, wavelets, Notices Amer. Math. Soc. 50 (2003), 880–894.
49. P. E. T. Jorgensen, Measures in wavelet decompositions, preprint, University of Iowa, 2004, submitted to Adv. in Appl. Math., arXiv:math.CA/0402024.
50. P. E. T. Jorgensen, Iterated function systems, representations, and Hilbert space, preprint, University of Iowa, 2004, to appear in Internat. J. Math., arXiv:math.CA/0402175.
51. P. E. T. Jorgensen and S. Pedersen, Spectral theory for Borel sets in $\mathbb{R}^n$ of finite measure, J. Funct Anal. 107 (1992), 72–104.
52. P. E. T. Jorgensen and S. Pedersen, Dense analytic subspaces in fractal L^2-spaces, J. Analyse Math. 75 (1998), 185–228.
53. P. E. T. Jorgensen and S. Pedersen, Spectral pairs in Cartesian coordinates, J. Fourier Anal. Appl. 5 (1999), 285–302.
54. J.-P. Kahane and P.-G. Lemarié-Rieusset, Fourier Series and Wavelets, Studies in the Development of Modern Mathematics, vol. 3, Gordon and Breach, Luxembourg, 1995.
55. G. Kaiser, A Friendly Guide to Wavelets, Birkhäuser, Boston, 1994.
56. A. Klappenecker, Wavelets and wavelet packets on quantum computers, in Wavelet Applications in Signal and Image Processing VII (Denver, 1999), ed. M. A. Unser, A. Aldroubi, and A. F. Laine, Proceedings of SPIE, vol. 3813, Society of Photo-optical Instrumentation Engineers, Bellingham, WA, 1999, pp. 703–713.
57. J. C. Lagarias, J. A. Reeds, and Y. Wang, Orthonormal bases of exponentials for the n-cube, Duke Math. J. 103 (2000), 25–37.
58. M. Lang and P. N. Heller, The design of maximally smooth wavelets, in IEEE International Conference on Acoustics, Speech, and Signal Processing, 1996 (ICASSP-96), Institute of Electrical and Electronics Engineers, New York, 1996, vol. 3, pp. 1463–1466.
59. W. Lawton, S. L. Lee, and Z. Shen, Convergence of multidimensional cascade algorithm, Numer. Math. 78 (1998), 427–438.
60. L.-H. Lim, J. A. Packer, and K. F. Taylor, A direct integral decomposition of the wavelet representation, Proc. Amer. Math. Soc. 129 (2001), 3057–3067.
61. D. Marr, Vision: A Computational Investigation into the Human Representation and Processing of Visual Information, W.H. Freeman, San Francisco, 1982.
62. Y. Meyer, Wavelets: Algorithms & Applications, SIAM, Philadelphia, 1993, translated from the French and with a foreword by Robert D. Ryan.
63. Y. Meyer, Wavelets, Vibrations and Scalings, CRM Monograph Series, vol. 9, American Mathematical Society, Providence, 1998.
64. Y. Meyer, Wavelets and functions with bounded variation from image processing to pure mathematics, Atti Accad. Naz. Lincei Cl. Sci. Fis. Mat. Natur. Rend. Lincei (9) Mat. Appl. 11 (2000), 77–105, special issue: *Mathematics*

Towards the Third Millennium (Papers from the International Conference held in Rome, May 27–29, 1999).
65. C. A. Micchelli and Y. Xu, Using the matrix refinement equation for the construction of wavelets on invariant sets, Appl. Comput. Harmon. Anal. 1 (1994), 391–401.
66. A. M. Perelomov, Generalized Coherent States and Their Applications, Texts and Monographs in Physics, Springer-Verlag, Berlin, 1986.
67. V. Perrier and M. V. Wickerhauser, Multiplication of short wavelet series using connection coefficients, in Advances in Wavelets (Hong Kong, 1997), ed. K.-S. Lau, Springer-Verlag, Singapore, 1999, pp. 77–101.
68. H. L. Resnikoff and R. O. Wells, Wavelet Analysis: The Scalable Structure of Information, Springer-Verlag, New York, 1998.
69. S. D. Riemenschneider and Z. Shen, Box splines, cardinal series, and wavelets, in Approximation Theory and Functional Analysis (College Station, Texas, 1990), ed. C. K. Chui, Academic Press, Boston, 1991, pp. 133–149.
70. S. D. Riemenschneider and Z. Shen, Wavelets and pre-wavelets in low dimensions, J. Approx. Theory 71 (1992), 18–38.
71. S. D. Riemenschneider and Z. Shen, Interpolatory wavelet packets, Appl. Comput. Harmon. Anal. 8 (2000), 320–324.
72. A. Ron and Z. Shen, Affine systems in $L_2(R^d)$: the analysis of the analysis operator, J. Funct. Anal. 148 (1997), 408–447.
73. A. Ron and Z. Shen, Compactly supported tight affine spline frames in $L_2(R^d)$, Math. Comp. 67 (1998), 191–207.
74. A. Ron and Z. Shen, The Sobolev regularity of refinable functions, J. Approx. Theory 106 (2000), 185–225.
75. A. Ron and Z. Shen, The wavelet dimension function is the trace function of a shift-invariant system, Proc. Amer. Math. Soc. 131 (2003), 1385–1398.
76. A. Ron, Z. Shen, and K.-C. Toh, Computing the Sobolev regularity of refinable functions by the Arnoldi method, SIAM J. Matrix Anal. Appl. 23 (2001), 57–76.
77. M. Rørdam, F. Larsen, and N. J. Laustsen, An Introduction to K-theory for C^*-algebras, Cambridge University Press, Cambridge, 2000.
78. Z. Shen, Refinable function vectors, SIAM J. Math. Anal. 29 (1998), 235–250.
79. G. Strang, Wavelet transforms versus Fourier transforms, Bull. Amer. Math. Soc. (N.S.) 28 (1993), 288–305.
80. G. Strang and T. Nguyen, Wavelets and Filter Banks, Wellesley-Cambridge Press, Wellesley, Massachusetts, 1996.
81. G. Strang, V. Strela, and D.-X. Zhou, Compactly supported refinable functions with infinite masks, in The Functional and Harmonic Analysis of Wavelets and Frames (San Antonio, 1999), ed. L. W. Baggett and D. R. Larson, Contemp. Math., vol. 247, American Mathematical Society, Providence, 1999, pp. 285–296.
82. G. Strang and D.-X. Zhou, Inhomogeneous refinement equations, J. Fourier Anal. Appl. 4 (1998), 733–747.
83. G. Strang and D.-X. Zhou, The limits of refinable functions, Trans. Amer. Math. Soc. 353 (2001), 1971–1984.

84. V. Strela, P. N. Heller, G. Strang, P. Topiwala, and C. Heil, The application of multiwavelet filterbanks to image processing, IEEE Transactions on Image Processing 8 (1999), 548–563.
85. T. Tao, Fuglede's conjecture is false in 5 and higher dimensions, Math. Res. Lett. 11 (2004), 251–258.
86. B. F. Treadway, Appendix to [46].
87. P. P. Vaidyanathan, Multirate Systems and Filter Banks, Prentice Hall, Englewood Cliffs, NJ, 1993.
88. M. V. Wickerhauser, Best-adapted wavelet packet bases, in Different Perspectives on Wavelets (San Antonio, TX, 1993), ed. I. Daubechies, Proc. Sympos. Appl. Math., vol. 47, American Mathematical Society, Providence, 1993, pp. 155–171.

UNITARY SYSTEMS, WAVELET SETS, AND OPERATOR-THEORETIC INTERPOLATION OF WAVELETS AND FRAMES

David R. Larson

Department of Mathematics, Texas A&M University
College Station, TX 77845, U.S.A.
E-mail: larson@math.tamu.edu

A wavelet is a special case of a vector in a separable Hilbert space that generates a basis under the action of a collection, or system, of unitary operators. We will describe the operator-interpolation approach to wavelet theory using the local commutant of a system. This is really an abstract application of the theory of operator algebras to wavelet theory. The concrete applications of this method include results obtained using specially constructed families of wavelet sets. A frame is a sequence of vectors in a Hilbert space which is a compression of a basis for a larger space. This is not the usual definition in the frame literature, but it is easily equivalent to the usual definition. Because of this compression relationship between frames and bases, the unitary system approach to wavelets (and more generally: wandering vectors) is perfectly adaptable to frame theory. The use of the local commutant is along the same lines as in the wavelet theory. Finally, we discuss constructions of frames with special properties using targeted decompositions of positive operators, and related problems.

1. Introduction

This is a write-up of a tutorial series of three talks which I gave as part of the "Workshop on Functional and Harmonic Analyses of Wavelets and Frames" held August 4-7, 2004 at the National University of Singapore. I will first give the titles and abstracts essentially as they appeared in the workshop schedule. I will say that the actual style of write-up of these notes will be structured a bit differently, but only in that more than three sections will be given, and subsections indicated, to (hopefully) improve expositional quality.

1.1. *Talks and abstracts*

(a) "Unitary Systems and Wavelet Sets": A wavelet is a special case of a vector in a separable Hilbert space that generates a basis under the action of a collection, or "system", of unitary operators defined in terms of translation and dilation operations. This approach to wavelet theory goes back, in particular, to earlier work of Goodman, Lee and Tang [25] in the context of multiresolution analysis. We will begin by describing the operator-interpolation approach to wavelet theory using the local commutant of a system that was worked out by the speaker and his collaborators a few years ago. This is really an abstract application of the theory of operator algebras, mainly von Neumann algebras, to wavelet theory. The concrete applications of operator-interpolation to wavelet theory include results obtained using specially constructed families of wavelet sets. In fact X. Dai and the speaker had originally developed our theory of wavelet sets [11] specifically to take advantage of their natural and elegant relationships with these wavelet unitary systems. We will also discuss some new results and open questions.

(b) "Unitary Systems and Frames": A frame is a sequence of vectors in a Hilbert space which is a compression of a basis for a larger space. (This is not the usual definition in the frame literature, but it is equivalent to the usual definition. In this spirit, the usual "inequality" definition can be thought of as an abstract characterization of a compression of a basis.) Because of this compression relationship between frames and bases, the unitary system approach to wavelets (and more generally: wandering vectors) is perfectly adaptable to frame theory. This idea was developed into a theory a few years ago by D. Han and the speaker [33]. The use of the local commutant is along the same lines.

(c) "Decompositions of Operators and Operator-Valued Frames": We will discuss some joint work with K. Kornelson and others on construction of frames with targeted properties [16, 42]. These are related to targeted decompositions of positive operators.

1.2. *Some background*

It might be appropriate to give some comments of a personal-historical nature, before continuing with the technical aspects. My particular point of view on "wavelet theory", which was developed jointly with my good friend and colleague Xingde Dai, began in the summer of 1992. Before then, I was strictly an operator theorist. I had heard my approximation theory colleagues and friends at Texas A&M University talk about wavelets and frames, and Xingde had frequently mentioned these topics to me when he was finishing his Ph.D. at A&M (he was my student, graduating in 1990, with a thesis [9] on the subject of *nest algebras*). But it was this meeting of minds we had in June of 1992 that was the turning point for me. We formulated an approach to wavelet theory (and as it turned out ultimately, to frame-wavelet theory as well) that I felt we could "really understand" as operator algebraists. This was the abstract *unitary system* approach. Dai knew the unitary operator approach to multiresolution analysis that had been recently (at that time) published by Goodman, Lee and Tang [25], and he suggested to me that we should try to go further with these ideas in an attempt to get some type of tractable classification of *all* wavelets. We went, in fact, in some completely different directions. The first paper that came out of this was the AMS Memoir [11] with Dai. The second paper was our paper [12] with Dai and Speegle, which proved the existence of single wavelets in higher dimensions, for arbitrary expansive dilations. After that, several papers followed including the AMS Memoir [33] with Deguang Han, and the Wutam Consortium paper [52], as well as the papers [10, 13, 26, 27, 28, 30, 34, 43, 44] by my students Dai, Gu, Han and Lu, and their collaborators, and the papers [2, 39, 40] with my colleagues Azoff, Ionascu, and Pearcy.

The paper [11] with Dai mentioned above, which was published in 1998, culminated about two years of work on this topic by the authors. It contained our entire operator-theoretic approach to wavelet theory, and was completed in December 1994. This work, while theoretical, had much hands-on experimentation in its development, and resulted in certain theorems we were able to prove concerning constructions of new families of wavelets. In order to conduct successful *experiments* with our operator techniques, we needed a supply of easily computable *test* wavelets: that is, wavelets which were very amenable to *paper and pencil computations*. We discovered that certain sets, we called *wavelet sets*, existed in abundance, and we computed many concrete examples of them along the way toward proving our results of [11]. Several of these were given as examples in [11, Example 4.5, items $(i) \rightarrow (xi)$]. Some of these are given in Section 2.6.1 of the present article.

Most of our work in [11] on the local commutant and the theory of wavelet sets was accomplished in the two-month period July-September 1992. The first time Dai and I used the terms *wavelet set* and *local commutant*, as well as the first time we discussed what we referred to as the *connectedness problem for wavelets*, was in a talk in a Special Session on

Operator Algebras in the October 1992 AMS Sectional Meeting in Dayton.

Along the way a graduate student at A&M, Darrin Speegle, who was enrolled in a seminar course of Larson on the manuscript of [11], answered an open question Larson gave out in class by proving that the set of all wavelet sets for a given wavelet system is *connected* in the symmetric difference metric on the class of measurable sets of finite measure. That resulted in a paper [50] which became part of his thesis (which was directed by William Johnson of Texas A&M), and Speegle subsequently joined forces with Dai and Larson [12] to prove that wavelet sets (and indeed, wavelets) exist in much greater generality than the prevailing *folklore* dictated. We received some attention for our work, and especially we thank Guido Weiss and John Benedetto for recognizing our work. This led to a flurry of papers by a number of authors, notably [4, 6, 13], and also led to the paper [52] by the Wutam Consortium, which was a group led by Guido Weiss and Larson, consisting of 14 researchers–students and postdocs of Weiss and Larson–based at Washington University and Texas A&M University, for the purpose of doing basic research on wavelet theory.

1.2.1. *Interpolation*

The main point of the *operator-theoretic interpolation* of wavelets (and frames) that Dai and I developed is that new wavelets can be obtained as linear combinations of known ones using *coefficients* which are not necessarily scalars but can be taken to be *operators* (in fact, *Fourier multipliers*) in a certain class. The ideas involved in this, and the essential computations, all extend naturally to more general unitary systems and *wandering vectors*, and I think that much of the theory is best-put in this abstract setting because clarity is enhanced, and because many of the methods work for more involved systems that are important to applied harmonic analysis, such as Gabor and generalized Gabor systems, and various types of *frame* unitary systems.

1.2.2. *Some basic terminology*

This article will concern bounded linear operators on separable Hilbert spaces. The set of all bounded linear operators on a Hilbert space H will be denoted by $B(H)$. By a *bilateral shift* U on H we mean a unitary operator U for which there exists a closed linear subspace $E \subset H$ with the property that the family of subspaces $\{U^n E : n \in \mathbb{Z}\}$ are orthogonal and give a direct-sum decomposition of H. The subspace E is called a *complete wandering subspace* for U. The *multiplicity* of U is defined to be the dimension of E.

The *strong operator topology* on $B(H)$ is the topology of pointwise convergence, and the *weak operator topology* is the weakest topology such that the vector functionals $\omega_{x,y}$ on $B(H)$ defined by $A \mapsto \langle Ax, y\rangle$, $A \in B(H)$, $x, y \in H$, are all continuous. An *algebra of operators* is a linear subspace of $B(H)$ which is closed under multiplication. An *operator algebra* is an algebra of operators which is *norm-closed.* A subset $\mathcal{S} \subset B(H)$ is called *selfadjoint* if whenever $A \in \mathcal{S}$ then also $A^* \in \mathcal{S}$. A *C^*-algebra* is a selfadjoint operator algebra. A *von Neumann algebra* is a C^*-algebra which is closed in the weak operator topology. For a unital operator algebra, it is well known that being closed in the weak operator topology is equivalent to being closed in the strong operator topology.

The *commutant* of a set $\mathcal{S}$ of operators in $B(H)$ is the family of all operators in $B(H)$ that *commute* with every operator in $\mathcal{S}$. It is closed under addition and multiplication, so is an algebra. And it is clearly closed in both the weak operator topology and the strong operator topology. We use the standard *prime* notation for the commutant. So the commutant of a subset $\mathcal{S} \subset B(H)$ is denoted: $\mathcal{S}' := \{A \in B(H) : AS = SA, \quad S \in \mathcal{S}\}$.

The commutant of a selfadjoint set of operators is clearly a von Neumann algebra. Moreover, by a famous theorem of Fuglede every operator which commutes with a normal operator N also commutes with its adjoint N^*, and hence the commutant of any set of *normal* operators is also a von Neumann algebra. So, of particular relevance to this work, the commutant of any set of *unitary* operators is a von Neumann algebra.

One of the main tools in this work is the *local commutant* of a system of unitary operators. (See section 2.4.) This is a natural generalization of the commutant of the system, and like the commutant it is a linear space of operators which is closed in the weak and the strong operator topologies, but unlike the commutant it is usually not selfadjoint, and is usually not closed under multiplication. It contains the commutant of the system, but can be much larger than the commutant. The local commutant of a wavelet unitary system captures all the information about the wavelet system in an essential way, and this gives the *flavor* of our approach to the subject.

If U is a unitary operator and $\mathcal{A}$ is an operator algebra, then U is said to *normalize* $\mathcal{A}$ if $U^\star \cdot \mathcal{A} \cdot U = \mathcal{A}$. In the most interesting cases of operator-theoretic interpolation: that is, those cases that yield the strongest structural results, the relevant unitaries in the local commutant of the system normalize the commutant of the system.

1.2.3. *Acknowledgements*

I want to take the opportunity to thank the organizers of this wonderful workshop at the National University of Singapore for their splendid hospitality and great organization, and for inviting me to give the series of tutorial-style talks that resulted in this write-up. I also want to state that the work discussed in this article was supported by grants from the United States National Science Foundation.

2. Unitary Systems and Wavelet Sets

We define a *unitary system* to be simply a collection of unitary operators $\mathcal{U}$ acting on a Hilbert space H which contains the identity operator. The *interesting* unitary systems all have additional structural properties of various types. We will say that a vector $\psi \in H$ is *wandering* for $\mathcal{U}$ if the set

$$\mathcal{U}\psi := \{U\psi : U \in \mathcal{U}\} \tag{1}$$

is an orthonormal set, and we will call ψ a *complete wandering vector* for $\mathcal{U}$ if $\mathcal{U}\psi$ spans H. This (abstract) point of view can be useful. Write

$$\mathcal{W}(\mathcal{U})$$

for the set of complete wandering vectors for $\mathcal{U}$.

2.1. *The one-dimensional wavelet system*

For simplicity of presentation, much of the work in this article will deal with one-dimensional wavelets, and in particular, the dyadic case. The other cases: non-dyadic and in higher dimensions, are well-described in the literature and are at least notationally more complicated.

2.1.1. *Dyadic wavelets*

A *dyadic orthonormal* wavelet in one dimension is a unit vector $\psi \in L^2(\mathbb{R}, \mu)$, with μ Lebesgue measure, with the property that the set

$$\{2^{\frac{n}{2}}\psi(2^n t - l) : n, l \in \mathbb{Z}\} \tag{2}$$

of all integral translates of ψ followed by dilations by arbitrary integral powers of 2, is an orthonormal basis for $L^2(\mathbb{R}, \mu)$. The term *dyadic* refers to the dilation factor "2". The term *mother wavelet* is also used in the literature for ψ. Then the functions

$$\psi_{n,l} := 2^{\frac{n}{2}} \psi(2^n t - l)$$

are called elements of the wavelet basis generated by the "mother". The functions $\psi_{n,l}$ will not themselves be mother wavelets unless $n = 0$.

Let T and D be the translation (by 1) and dilation (by 2) unitary operators in $B(L^2(\mathbb{R})$ given by $(Tf)(t) = f(t-1)$ and $(Df)(t) = \sqrt{2}f(2t)$. Then

$$2^{\frac{n}{2}} \psi(2^n t - l) = (D^n T^l \psi)(t)$$

for all $n, l \in \mathbb{Z}$. Operator-theoretically, the operators T, D are *bilateral shifts* of *infinite multiplicity*. It is obvious that $L^2([0,1])$, considered as a subspace of $L^2(\mathbb{R})$, is a complete wandering subspace for T, and that $L^2([-2,-1] \cup [1,2])$ is a complete wandering subspace for D.

2.1.2. *The dyadic unitary system*

Let $\mathcal{U}_{D,T}$ be the unitary system defined by

$$\mathcal{U}_{D,T} = \{D^n T^l : n, l \in \mathbb{Z}\} \tag{3}$$

where D and T are the operators defined above. Then ψ is a dyadic orthonormal wavelet if and only if ψ is a complete wandering vector for the unitary system $\mathcal{U}_{D,T}$. This was our original motivation for developing the abstract unitary system theory. Write

$$\mathcal{W}(D,T) := \mathcal{W}(\mathcal{U}_{D,T}) \tag{4}$$

to denote the set of all dyadic orthonormal wavelets in one dimension.

An abstract interpretation is that, since D is a bilateral shift it has (many) complete wandering subspaces, and a wavelet for the system is a vector ψ whose translation space (that is, the closed linear span of $\{T^k : k \in \mathbb{Z}\}$ is a complete wandering subspace for D. Hence ψ must generate an orthonormal basis for the entire Hilbert space under the action of the unitary system.

2.1.3. *Non-dyadic wavelets in one dimension*

In one dimension, there are non-dyadic orthonormal wavelets: i.e. wavelets for all possible dilation factors besides 2 (the dyadic case). We said "possible", because the scales $\{0, 1, -1\}$ are excluded as scales because the dilation operators they would introduce are not bilateral shifts. All other real numbers for scales yield wavelet theories. In [11, Example 4.5 (x)] a family of examples is given of three-interval wavelet sets (and hence wavelets) for all scales $d \geq 2$, and it was noted there that such a family also exists for dilation factors $1 < d \leq 2$. There is some recent (yet unpublished) work that has been done, by REU students and mentors, building on this, classifying finite-interval wavelet sets for all possible real (positive and negative scale factors). I mentioned this work, in passing, in my talk.

2.2. *N dimensions*

2.2.1. *The expansive-dilation case*

Let $1 \leq m < \infty$, and let A be an $n \times n$ real matrix which is *expansive* (equivalently, all (complex) eigenvalues have modulus > 1). By a *dilation - A regular-translation orthonormal wavelet* we mean a function $\psi \in L^2(\mathbb{R}^n)$ such that

$$\{|det(A)|^{\frac{n}{2}}\psi(A^n t - (l_1, l_2, ..., l_n)^t) : n, l \in \mathbb{Z}\} \tag{5}$$

where $t = (t_1, ..., t_n)^t$, is an orthonormal basis for $L^2(\mathbb{R}^n; m)$. (Here m is product Lebesgue measure, and the superscript "t" means transpose.)

If $A \in M_n(\mathbb{R})$ is invertible (so in particular if A is expansive), then it is very easy to verify that the operator defined by

$$(D_A f)(t) = |det A|^{\frac{1}{2}} f(At) \tag{6}$$

for $f \in L^2(\mathbb{R}^n)$, $t \in \mathbb{R}^n$, is *unitary.* For $1 \leq i \leq n$, let T_i be the unitary operator determined by translation by 1 in the ith coordinate direction. The set (5) above is then

$$\{D_A^k T_1^{l_1} \cdots T_n^{l_n} \psi : k, l_i \in \mathbb{Z}\} \tag{7}$$

If the dilation matrix A is expansive, but the translations are along some oblique lattice, then there is an invertible real $n \times n$ matrix T such that conjugation with D_T takes the entire wavelet system to a regular-translation expansive-dilation matrix. This is easily worked out, and was shown in detail in [39] in the context of working out a complete theory of unitary equivalence of wavelet systems. Hence the wavelet theories are equivalent.

2.2.2. *The non-expansive dilation case*

Much work has been accomplished concerning the existence of wavelets for dilation matrices A which are not expansive. Some of the original work was accomplished in the Ph.D. theses of Q. Gu and D. Speegle, when they were together finishing up at Texas A&M. Some significant additional work was accomplished by Speegle in [49], and also by others. In [39], with Ionascu and Pearcy we proved that if an $n \times n$ real invertible matrix A is not similar (in the $n \times n$ complex matrices) to a unitary matrix, then the corresponding dilation operator D_A is in fact a bilateral shift of infinite multiplicity. If a dilation matrix were to admit any type of wavelet (or frame-wavelet) theory, then it is well-known that a necessary condition would be that the corresponding dilation operator would have to be a bilateral shift of infinite multiplicity. I am happy to report that in very recent work [45], with E. Schulz, D. Speegle, and K. Taylor, we have succeeded in showing that this minimal condition is in fact sufficient: such a matrix, with regular translation lattice, admits a (perhaps infinite) tuple of functions, which collectively generates a frame-wavelet under the action of this unitary system.

2.3. *Abstract systems*

2.3.1. *Restrictions on wandering vectors*

We note that *most* unitary systems $\mathcal{U}$ do not have complete wandering vectors. For $\mathcal{W}(\mathcal{U})$ to be nonempty, the set $\mathcal{U}$ must be very special. It must be *countable* if it acts separably (i.e. on a separable Hilbert space), and it must be *discrete* in the strong operator topology because if $U, V \in \mathcal{U}$ and if x is a wandering vector for $\mathcal{U}$ then

$$\|U - V\| \geq \|Ux - Vx\| = \sqrt{2}$$

Certain other properties are forced on $\mathcal{U}$ by the presence of a wandering vector. One purpose of [11] was to study such properties. Indeed, it was a matter of some surprise to us to discover that such a theory is viable even in some considerable generality. For perspective, it is useful to note that while $\mathcal{U}_{D,T}$ has complete wandering vectors, the reversed system

$$\mathcal{U}_{T,D} = \{T^l D^n : n, l \in \mathbb{Z}\}$$

fails to have a complete wandering vector. (A proof of this was given in the introduction to [11].)

2.3.2. *Group systems*

An example which is important to the theory is the following: let G be an arbitrary countable group, and let $H = l^2(G)$. Let π be the (left) regular representation of G on H. Then every element of G gives a complete wandering vector for the unitary system

$$\mathcal{U} := \pi(G).$$

(If $h \in G$ it is clear that the vector $\lambda_h \in l^2(G)$, which is defined to have 1 in the h position and 0 elsewhere, is in $\mathcal{W}(\mathcal{U})$.) If a unitary system is a *group*, and if it has a complete wandering vector, it is not hard to show that it is unitarily equivalent to this example.

2.4. *The local commutant*

2.4.1. *The local commutant of the system $\mathcal{U}_{D,T}$*

Computational aspects of operator theory can be introduced into the wavelet framework in an elementary way. Here is the way we originally did it: Fix a wavelet ψ and consider the set of all operators $S \in B(L^2(\mathbb{R}))$ which *commute* with the *action* of dilation and translation on ψ. That is, require

$$(S\psi)(2^n t - l) = S(\psi(2^n t - l)) \tag{8}$$

or equivalently

$$D^n T^l S\psi = SD^n T^l \psi \tag{9}$$

for all $n, l \in \mathbb{Z}$. Call this the *local commutant of the wavelet system $\mathcal{U}_{D,T}$ at the vector ψ*. (In our first preliminary writings and talks we called it the *point commutant* of the system.) Formally, the local commutant of the dyadic wavelet system on $L^2(\mathbb{R})$ is:

$$\mathcal{C}_\psi(\mathcal{U}_{D,T}) := \{S \in B(L^2(\mathbb{R})) : (SD^nT^l - D^nT^lS)\psi = 0, \forall n, l \in \mathbb{Z}\} \tag{10}$$

This is a linear subspace of $B(H)$ which is closed in the strong operator topology, and in the weak operator topology, and it clearly contains the *commutant* of $\{D, T\}$.

A motivating example is that if η is any other wavelet, let $V := V_\psi^\eta$ be the unitary (we call it the *interpolation unitary*) that takes the basis $\psi_{n,l}$ to the basis $\eta_{n,l}$. That is, $V\psi_{n,l} = \eta_{n,l}$ for all $n, l \in \mathbb{Z}$. Then $\eta = V\psi$, so $VD^nT^l\psi = D^nT^lS\psi$ hence $V \in \mathcal{C}_\psi(\mathcal{U}_{D,T})$.

In the case of a pair of complete wandering vectors ψ, η for a general unitary system $\mathcal{U}$, we will use the same notation V_ψ^η for the unitary that takes the vector $U\psi$ to $U\eta$ for all $U \in \mathcal{U}$.

This simple-minded idea is reversible, so for every unitary V in $\mathcal{C}_\psi(\mathcal{U}_{D,T})$ the vector $V\psi$ is a wavelet. This correspondence between unitaries in

$\mathcal{C}_\psi(D,T)$ and dyadic orthonormal wavelets is one-to-one and onto (see Proposition 1). This turns out to be useful, because it leads to some new formulas relating to decomposition and factorization results for wavelets, making use of the *linear* and *multiplicative* properties of $\mathcal{C}_\psi(D,T)$.

It turns out (a proof is required) that the entire local commutant of the system $\mathcal{U}_{D,T}$ at a wavelet ψ is *not* closed under multiplication, but it also turns out (also via a proof) that for *most* (and perhaps *all*) wavelets ψ the local commutant at ψ contains many noncommutative operator algebras (in fact von Neumann algebras) as subsets, and their unitary groups *parameterize* norm-arcwise-connected families of wavelets. Moreover, $\mathcal{C}_\psi(D,T)$ is closed under *left multiplication* by the commutant $\{D,T\}'$, which turns out to be an abelian nonatomic von Neumann algebra. The fact that $\mathcal{C}_\psi(D,T)$ is a *left module* under $\{D,T\}'$ leads to a method of obtaining new wavelets from old, and of obtaining connectedness results for wavelets, which we called *operator-theoretic interpolation* of wavelets in [11], (or simply *operator-interpolation*).

2.4.2. *The local commutant of an abstract unitary system*

More generally, let $\mathcal{S} \subset B(H)$ be a set of operators, where H is a separable Hilbert space, and let $x \in H$ be a nonzero vector, and *formally* define the *local commutant* of $\mathcal{S}$ at x by

$$\mathcal{C}_x(\mathcal{S}) := \{A \in B(H) : (AS - SA)x = 0, S \in \mathcal{S}\}$$

As in the wavelet case, this is a weakly and strongly closed linear subspace of $B(H)$ which contains the commutant $\mathcal{S}'$ of $\mathcal{S}$. If x is *cyclic* for $\mathcal{S}$ in the sense that span$(\mathcal{S}x)$ is dense in H, then x *separates* $\mathcal{C}_x(\mathcal{S})$ in the sense that for $S \in \mathcal{C}_x(\mathcal{S})$, we have $Sx = 0$ iff $x = 0$. Indeed, if $A \in \mathcal{C}_x(\mathcal{S})$ and if $Ax = 0$, then for any $S \in \mathcal{S}$ we have $ASx = SAx = 0$, so $ASx = 0$, and hence $A = 0$.

If $A \in \mathcal{C}_x(\mathcal{S})$ and $B \in \mathcal{S}'$, let $C = BA$. Then for all $S \in \mathcal{S}$,

$$(CS - SC)x = B(AS)x - (SB)Ax = B(SA)x - (BS)Ax = 0$$

because $ASx = SAx$ since $A \in \mathcal{C}_x(\mathcal{S})$, and $SB = BS$ since $B \in \mathcal{S}'$. Hence $\mathcal{C}_x(\mathcal{S})$ is closed under left multiplication by operators in $\mathcal{S}'$. That is, $\mathcal{C}_x(\mathcal{S})$ is a *left module* over $\mathcal{S}'$.

It is interesting that, if in addition $\mathcal{S}$ is a multiplicative semigroup, then in fact $\mathcal{C}_x(\mathcal{S})$ is identical with the commutant $\mathcal{S}'$ so in this case the commutant is not a new structure. To see this, suppose $A \in \mathcal{C}_x(\mathcal{S})$. Then for each $S, T \in \mathcal{S}$ we have $ST \in \mathcal{S}$, and so

$$AS(Tx) = (ST)Ax = S(ATx) = (S)Tx$$

So since $T \in \mathcal{S}$ was arbitrary and $\text{span}(\mathcal{S}x) = H$, it follows that $AS = SA$.

Proposition 1: *If $\mathcal{U}$ is any unitary system for which $\mathcal{W}(\mathcal{U}) \neq \emptyset$, then for any $\psi \in \mathcal{W}(\mathcal{U})$*

$$\mathcal{W}(\mathcal{U}) = \{U\psi : U \text{ is a unitary operator in } \mathcal{C}_\psi(\mathcal{U})\}$$

and the correspondence $U \to U\psi$ is one-to-one.

A *Riesz basis* for a Hilbert space H is the image under a bounded invertible operator of an orthonormal basis. Proposition 1 generalizes to generators of Riesz bases. A *Riesz vector* for a unitary system $\mathcal{U}$ is defined to be a vector ψ for which $\mathcal{U}\psi := \{U\psi : U \in \mathcal{U}\}$ is a Riesz basis for the closed linear span of $\mathcal{U}\psi$, and it is called *complete* if $\overline{\text{span}}\, \mathcal{U}\psi = H$. Let $\mathcal{RW}(\mathcal{U})$ denote the set of all complete Riesz vectors for $\mathcal{U}$.

Proposition 2: *Let $\mathcal{U}$ be a unitary system on a Hilbert space H. If ψ is a complete Riesz vector for $\mathcal{U}$, then*

$$\mathcal{RW}(\mathcal{U}) = \{A\psi : A \text{ is an operator in } \mathcal{C}_\psi(\mathcal{U}) \text{ that is invertible in } B(H)\}.$$

2.4.3. *Operator-theoretic interpolation*

Now suppose $\mathcal{U}$ is a unitary system, such as $\mathcal{U}_{D,T}$, and suppose $\{\psi_1, \psi_2, \ldots, \psi_m\} \subset \mathcal{W}(\mathcal{U})$. (In the case of $\mathcal{U}_{D,T}$, this means that $(\psi_1, \psi_2, \ldots, \psi_n)$ is an n-tuple of wavelets.)

Let $(A_1, A_2, \ldots, A_n)$ be an n-tuple of operators in the commutant $\mathcal{U}'$ of $\mathcal{U}$, and let η be the vector

$$\eta := A_1\psi_1 + A_2\psi_2 + \cdots + A_n\psi_n \ .$$

Then

$$\begin{aligned}\eta &= A_1\psi_1 + A_2 V_{\psi_1}^{\psi_2}\psi_1 + \ldots A_n V_{\psi_1}^{\psi_n}\psi_1 \\ &= (A_1 + A_2 V_{\psi_1}^{\psi_2} + \cdots + A_n V_{\psi_1}^{\psi_n})\psi_1 \ . \qquad (11)\end{aligned}$$

We say that η is obtained by *operator interpolation* from $\{\psi_1, \psi_2, \ldots, \psi_m\}$. Since $\mathcal{C}_{\psi_1}(\mathcal{U})$ is a left $\mathcal{U}'$ - module, it follows that the operator

$$A := A_1 + A_2 V_{\psi_1}^{\psi_2} + \cdots + A_n V_{\psi_1}^{\psi_n} \qquad (12)$$

is an element of $\mathcal{C}_{\psi_1}(\mathcal{U})$. Moreover, if B is another element of $\mathcal{C}_{\psi_1}(\mathcal{U})$ such that $\eta = B\psi_1$, then $A - B \in \mathcal{C}_{\psi_1}(\mathcal{U})$ and $(A - B)\psi_1 = 0$. So since ψ_1 *separates* $\mathcal{C}_{\psi_1}(\mathcal{U})$ it follows that $A = B$. Thus A is the *unique* element of $\mathcal{C}_{\psi_1}(\mathcal{U})$ that takes ψ_1 to η. Let $\mathcal{S}_{\psi_1,\ldots,\psi_n}$ be the family of all finite sums of the form

$$\sum_{i=0}^{n} A_i V_{\psi_1}^{\psi_i} \quad .$$

This is the left module of $\mathcal{U}'$ generated by $\{I, V_{\psi_1}^{\psi_2}, \ldots, V_{\psi_1}^{\psi_n}\}$. It is the $\mathcal{U}'$-*linear span* of $\{I, V_{\psi_1}^{\psi_2}, \ldots, V_{\psi_1}^{\psi_n}\}$. Let

$$\mathcal{M}_{\psi_1,\ldots,\psi_n} := (\mathcal{S}_{\psi_1,\ldots,\psi_n})\psi_1 \tag{13}$$

So

$$\mathcal{M}_{\psi_1,\ldots,\psi_n} = \left\{ \sum_{i=0}^{n} A_i \psi_i \ : \ A_i \in \mathcal{U}' \right\} \quad .$$

We call this the *interpolation space* for $\mathcal{U}$ generated by $(\psi_1, \ldots, \psi_n)$. From the above discussion, it follows that for every vector $\eta \in \mathcal{M}_{\psi_1,\psi_2,\ldots,\psi_n}$ there exists a unique operator $A \in \mathcal{C}_{\psi_1}(\mathcal{U})$ such that $\eta = A\psi_1$, and moreover this A is an element of $\mathcal{S}_{\psi_1,\ldots,\psi_n}$.

2.4.4. *Normalizing the commutant*

In certain essential cases (and we are not sure how general this type of case is) one can prove that an interpolation unitary V_ψ^η *normalizes* the commutant $\mathcal{U}'$ of the system in the sense that $V_\eta^\psi \mathcal{U}' V_\psi^\eta = \mathcal{U}'$. (Here, it is easily seen that $(V_\psi^\eta)^* = V_\eta^\psi$.) Write $V := V_\psi^\eta$. If V normalizes $\mathcal{U}'$, then the algebra, before norm closure, generated by $\mathcal{U}'$ and V is the set of all finite sums (trig polynomials) of the form $\sum A_n V^n$, with coefficients $A_n \in \mathcal{U}'$, $n \in \mathbb{Z}$. The closure in the strong operator topology is a von Neumann algebra. Now suppose further that *every power* of V is contained in $\mathcal{C}_\psi(\mathcal{U})$. This occurs only in special cases, yet it occurs frequently enough to yield some general methods. Then since $\mathcal{C}_\psi(\mathcal{U})$ is a SOT-closed linear subspace which is closed under left multiplication by $\mathcal{U}'$, this von Neumann algebra is contained in $\mathcal{C}_\psi(\mathcal{U})$, so its unitary group parameterizes a norm-path-connected subset of $\mathcal{W}(\mathcal{U})$ that contains ψ and η via the correspondence $U \to U\psi$.

In the special case of *wavelets*, this is the basis for the work that Dai and I did in [11, Chapter 5] on operator-theoretic interpolation of wavelets.

In fact, we specialized there and *reserved* the term *operator-theoretic interpolation* to refer explicitly to the case when the interpolation unitaries normalize the commutant. In some subsequent work, we *loosened* this restriction yielding our more general definition given in this article, because there are cases of interest in which we weren't able to prove normalization. However, it turns out that if ψ and η are s-elementary wavelets (see section 2.5.4), then indeed V_ψ^η normalizes $\{D,T\}'$. (See Proposition 14.) Moreover, V_ψ^η has a very special form: after conjugating with the Fourier transform, it is a composition operator with symbol a natural and very computable measure-preserving transformation of $\mathbb{R}$. In fact, it is precisely this special form for V_ψ^η that allows us to make the computation that it normalizes $\{D,T\}'$. On the other hand, we know of no pair (ψ,η) of wavelets for which V_ψ^η fails to normalize $\{D,T\}'$. The difficulty is simply that in general it is very hard to do the computations.

Problem: If $\{\psi,\eta\}$ is a pair of dyadic orthonormal wavelets, does the interpolation unitary V_ψ^η normalize $\{D,T\}'$? As mentioned above, the answer is yes if ψ and η are s-elementary wavelets.

2.4.5. *An elementary interpolation result*

The following result is the most elementary case of operator-theoretic interpolation.

Proposition 3: *Let $\mathcal{U}$ be a unitary system on a Hilbert space H. If ψ_1 and ψ_2 are in $\mathcal{W}(\mathcal{U})$, then*

$$\psi_1 + \lambda\psi_2 \in \mathcal{RW}(\mathcal{U})$$

for all complex scalars λ with $|\lambda| \neq 1$. More generally, if ψ_1 and ψ_2 are in $\mathcal{RW}(\mathcal{U})$ then there are positive constants $b > a > 0$ such that $\psi_1 + \lambda\psi_2 \in \mathcal{RW}(\mathcal{U})$ for all $\lambda \in \mathbb{C}$ with either $|\lambda| < a$ or with $|\lambda| > b$.

Proof: If $\psi_1, \psi_2 \in \mathcal{W}(\mathcal{U})$, let V be the unique unitary in $\mathcal{C}_{\psi_2}(\mathcal{U})$ given by Proposition 1 such that $V\psi_2 = \psi_1$. Then

$$\psi_1 + \lambda\psi_2 = (V + \lambda I)\psi_2.$$

Since V is unitary, $(V + \lambda I)$ is an invertible element of $\mathcal{C}_{\psi_2}(\mathcal{U})$ if $|\lambda| \neq 1$, so the first conclusion follows from Proposition 2. Now assume $\psi_1, \psi_2 \in \mathcal{RW}(\mathcal{U})$. Let A be the unique invertible element of $\mathcal{C}_{\psi_2}(\mathcal{U})$ such that $A\psi_2 =$

ψ_1, and write $\psi_1 + \lambda\psi_2 = (A + \lambda I)\psi_2$. Since A is bounded and invertible there are $b > a > 0$ such that

$$\sigma(A) \subseteq \{z \in \mathbb{C} : a < |z| < b\}$$

where $\sigma(A)$ denotes the spectrum of A, and the same argument applies. □

2.4.6. *Interpolation pairs of wandering vectors*

In some cases where a pair ψ, η of vectors in $\mathcal{W}(\mathcal{U})$ are given it turns out that the unitary V in $\mathcal{C}_\psi(\mathcal{U})$ with $V\psi = \eta$ happens to be a *symmetry* (i.e. $V^2 = I$). Such pairs are called *interpolation pairs* of wandering vectors, and in the case where $\mathcal{U}$ is a wavelet system, they are called interpolation pairs of wavelets. Interpolation pairs are more prevalent in the theory, and in particular the wavelet theory, than one might expect. In this case (and in more complex generalizations of this) certain linear combinations of complete wandering vectors are themselves complete wandering vectors – not simply complete Riesz vectors.

Proposition 4: *Let $\mathcal{U}$ be a unitary system, let $\psi, \eta \in \mathcal{W}(\mathcal{U})$, and let V be the unique operator in $\mathcal{C}_\psi(\mathcal{U})$ with $V\psi = \eta$. Suppose*

$$V^2 = I.$$

Then

$$\cos\alpha \cdot \psi + i \sin\alpha \cdot \eta \in \mathcal{W}(\mathcal{U})$$

for all $0 \leq \alpha \leq 2\pi$.

The above result can be thought of as the *prototype* of our operator-theoretic interpolation results. It is the second most elementary case. More generally, the scalar α in Proposition 4 can be replaced with an appropriate *self-adjoint operator* in the commutant of $\mathcal{U}$. In the wavelet case, after conjugating with the Fourier transform, which is a unitary operator, this means that α can be replaced with a wide class of nonnegative dilation-periodic (see definition below) bounded measurable functions on $\mathbb{R}$.

2.4.7. *A test for interpolation pairs*

The following converse to Proposition 4 is typical of the type of computations encountered in some wandering vector proofs.

Proposition 5: *Let $\mathcal{U}$ be a unitary system, let $\psi, \eta \in \mathcal{W}(\mathcal{U})$, and let V be the unique unitary in $\mathcal{C}_\psi(\mathcal{U})$ with $V\psi = \eta$. Suppose for some $0 < \alpha < \frac{\pi}{2}$ the vector*

$$\rho := \cos\alpha \cdot \psi + i \sin\alpha \cdot \eta$$

is contained in $\mathcal{W}(\mathcal{U})$. Then

$$V^2 = I.$$

Proof: Since $\mathcal{U}\psi$ is a basis it will be enough to show that $VU_1\psi = V^\star U_1\psi$ for all $U_1 \in \mathcal{U}$. So it will suffice to prove that for all $U_1, U_2 \in \mathcal{U}$ we have

$$\langle VU_1\psi, U_2\psi\rangle = \langle V^\star U_1\psi, U_2\psi\rangle.$$

Using the fact that V locally commutes with $\mathcal{U}$ at ψ we have

$$\langle VU_1\psi, U_2\psi\rangle = \langle U_1V\psi, U_2\psi\rangle = \langle U_1\eta, U_2\psi\rangle \qquad \text{and}$$

$$\langle V^\star U_1\psi, U_2\psi\rangle = \langle U_1\psi, VU_2\psi\rangle = \langle U_1\psi, U_2V\psi\rangle = \langle U_1\psi, U_2\eta\rangle.$$

So we must show that $\langle U_1\eta, U_2\psi\rangle = \langle U_1\psi, U_2\eta\rangle$ for all $U_1, U_2 \in \mathcal{U}$.
Write $\rho := \rho_\alpha$. By hypothesis ψ, η and ρ are unit vectors. So compute

$$1 = \langle \rho, \rho\rangle = \cos^2\alpha \cdot \langle\psi,\psi\rangle + i\sin\alpha\cos\alpha\cdot\langle\eta,\psi\rangle$$

$$- i\sin\alpha\cos\alpha\cdot\langle\psi,\eta\rangle + \sin^2\alpha\cdot\langle\eta,\eta\rangle$$

$$= 1 + i\sin\alpha\cos\alpha\cdot(\langle\eta,\psi\rangle - \langle\psi,\eta\rangle).$$

Thus, since $\sin\alpha\cos\alpha \neq 0$, we must have $\langle\eta,\psi\rangle = \langle\psi,\eta\rangle$. Also, for $U_1, U_2 \in \mathcal{U}$ with $U_1 \neq U_2$ we have

$$0 = \langle U_1\rho, U_2\rho\rangle = \cos^2\alpha\cdot\langle U_1\psi, U_2\psi\rangle + i\sin\alpha\cos\alpha\cdot\langle U_1\eta, U_2\psi\rangle$$

$$-i\sin\alpha\cos\alpha\cdot\langle U_1\psi, U_2\eta\rangle + \sin^2\alpha\cdot\langle U_1\eta, U_2\eta\rangle$$

$$= i\sin\alpha\cos\alpha\cdot(\langle U_1\eta, U_2\psi\rangle - \langle U_1\psi, U_2\eta\rangle),$$

which implies $\langle U_1\eta, U_2\psi\rangle = \langle U_1\psi, U_2\eta\rangle$ as required. □

The above result gives an *experimental method* of checking whether $V^2 = I$ for a given pair $\psi, \eta \in \mathcal{W}(\mathcal{U})$. One just checks whether

$$\rho := \frac{1}{\sqrt{2}}\psi + \frac{i}{\sqrt{2}}\eta$$

is an element of $\mathcal{W}(\mathcal{U})$, which is much simpler than attempting to work with the infinite matrix of V with respect to the basis $\mathcal{U}\psi$ (or some other basis for H).

2.4.8. *Connectedness*

If we consider again the example of the left regular representation π of a group G on $H := l^2(G)$, then the local commutant of $\mathcal{U} := \pi(G)$ at a vector $\psi \in \mathcal{W}(\pi(G))$ is just the commutant of $\pi(G)$. So since the unitary group of the von Neumann algebra $(\pi(G))'$ is norm-arcwise-connected, it follows that $\mathcal{W}(\pi(G))$ is norm-arcwise-connected.

Problem A in [11] asked whether $\mathcal{W}(D, L)$ is norm-arcwise-connected. It turned out that this conjecture was also formulated independently by Guido Weiss ([38], [37]) from a harmonic analysis point of view (our point of view was purely functional analysis), and this problem (and related problems) was the primary stimulation for the creation of the WUTAM CONSORTIUM – a team of 14 researchers based at Washington University and Texas A&M University. (See [52].)

This *connectedness conjecture* was answered yes in [52] for the special case of the family of dyadic orthonormal MRA wavelets in $L^2(\mathbb{R})$, but still remains open for the family of *arbitrary* dyadic orthonormal wavelets in $L^2(\mathbb{R})$.

In the wavelet case $\mathcal{U}_{D,T}$, if $\psi \in \mathcal{W}(D, T)$ then it turns out that $\mathcal{C}_\psi(\mathcal{U}_{D,T})$ is in fact *much larger* than $(\mathcal{U}_{D,T})' = \{D, T\}'$, underscoring the fact that $\mathcal{U}_{D,T}$ is NOT a group. In particular, $\{D, T\}'$ is abelian while $\mathcal{C}_\psi(\mathcal{D}, \mathcal{T})$ is nonabelian for every wavelet ψ. (The proof of these facts are contained in [11].)

2.5. *Wavelet sets*

2.5.1. *The Fourier transform*

We will use the following form of the Fourier–Plancherel transform $\mathcal{F}$ on $\mathcal{H} = L^2(\mathbb{R})$, because it is a form *normalized* so it is a unitary transformation. Although there is another such *normalized* form that is frequently used, and actually simpler, the present form is the one we used in our original first paper [11] involving operator theory and wavelets, and so we will stick with it in these notes to avoid any confusion to a reader of both.

If $f, g \in L^1(\mathbb{R}) \cap L^2(\mathbb{R})$ then

$$(\mathcal{F}f)(s) := \frac{1}{\sqrt{2\pi}} \int_{\mathbb{R}} e^{-ist} f(t)dt := \hat{f}(s), \tag{14}$$

and

$$(\mathcal{F}^{-1}g)(t) = \frac{1}{\sqrt{2\pi}} \int_{\mathbb{R}} e^{ist} g(s)ds. \tag{15}$$

We have

$$(\mathcal{F}T_\alpha f)(s) = \frac{1}{\sqrt{2\pi}} \int_{\mathbb{R}} e^{-ist} f(t-\alpha)dt = e^{-is\alpha}(\mathcal{F}f)(s).$$

So $\mathcal{F}T_\alpha\mathcal{F}^{-1}g = e^{-is\alpha}g$. For $A \in \mathcal{B}(\mathcal{H})$ let $\hat{A}$ denote $\mathcal{F}A\mathcal{F}^{-1}$. Thus

$$\widehat{T}_\alpha = M_{e^{-i\alpha s}}, \tag{16}$$

where for $h \in L^\infty$ we use M_h to denote the multiplication operator $f \to hf$. Since $\{M_{e^{-i\alpha s}}\colon\ \alpha \in \mathbb{R}\}$ generates the m.a.s.a. $\mathcal{D}(\mathbb{R}) := \{M_h\colon\ h \in L^\infty(\mathbb{R})\}$ as a von Neumann algebra, we have

$$\mathcal{F}\mathcal{A}_T\mathcal{F}^{-1} = \mathcal{D}(\mathbb{R}).$$

Similarly,

$$\begin{aligned}(\mathcal{F}D^n f)(s) &= \frac{1}{\sqrt{2\pi}} \int_{\mathbb{R}} e^{-ist}(\sqrt{2})^n f(2^n t)dt \\ &= (\sqrt{2})^{-n} \cdot \frac{1}{\sqrt{2\pi}} \int_{\mathbb{R}} e^{-i2^{-n}st} f(t)dt \\ &= (\sqrt{2})^{-2}(\mathcal{F}f)(2^{2^{-n}s}) = (D^{-n}\mathcal{F}f)(s).\end{aligned}$$

So $\widehat{D}^n = D^{-n} = D^{*n}$. Therefore,

$$\widehat{D} = D^{-1} = D^*. \tag{17}$$

2.5.2. *The commutant of* $\{D, T\}$

We have $\mathcal{F}\{D, T\}'\mathcal{F}^{-1} = \{\widehat{D}, \widehat{T}\}'$. It turns out that $\{\widehat{D}, \widehat{T}\}'$ has an easy characterization.

Theorem 6:

$$\{\widehat{D}, \widehat{T}\}' = \{M_h\colon\ h \in L^\infty(\mathbb{R}) \text{ and } h(s) = h(2s) \text{ a.e.}\}.$$

Proof: Since $\widehat{D} = D^*$ and D is unitary, it is clear that $M_h \in \{\widehat{D}, \widehat{T}\}'$ if and only if M_h commutes with D. So let $g \in L^2(\mathbb{R})$ be arbitrary. Then (a.e.) we have

$$\begin{aligned}(M_h Dg)(s) &= h(s)(\sqrt{2}\ g(2s)), \quad \text{and} \\ (DM_h g)(s) &= D(h(s)g(s)) = \sqrt{h}(2s)g(2s).\end{aligned}$$

Since these must be equal a.e. for arbitrary g, we must have $h(s) = h(2s)$ a.e. □

Now let $E = [-2,-1) \cup [1,2)$, and for $n \in \mathbb{Z}$ let $E_n = \{2^n x\colon\ x \in E\}$. Observe that the sets E_n are disjoint and have union $\mathbb{R}\backslash\{0\}$. So if g is any uniformly bounded function on E, then g extends uniquely (a.e.) to a function $\tilde{g} \in L^\infty(\mathbb{R})$ satisfying

$$\tilde{g}(s) = \tilde{g}(2s), \qquad s \in \mathbb{R},$$

by setting

$$\tilde{g}(2^n s) = g(s), \qquad s \in E, n \in \mathbb{Z},$$

and $\tilde{g}(0) = 0$. We have $\|\tilde{g}\|_\infty = \|g\|_\infty$. Conversely, if h is any function satisfying $h(s) = h(2s)$ a.e., then h is uniquely (a.e.) determined by its restriction to E. This 1-1 mapping $g \to M_{\tilde{g}}$ from $L^\infty(E)$ onto $\{\widehat{D},\widehat{T}\}'$ is a $*$-isomorphism.

We will refer to a function h satisfying $h(s) = h(2s)$ a.e. as a 2-*dilation periodic function.* This gives a simple algorithm for computing a large class of wavelets from a given one, by simply modifying the *phase*:

$$\begin{aligned}&\textit{Given } \psi, \textit{ let } \widehat{\psi} = \mathcal{F}(\psi), \textit{ choose a real-valued function } h \in L^\infty(E)\\ &\textit{arbitrarily, let } g = \exp(ih), \textit{ extend to a 2-dilation periodic}\\ &\textit{function } \tilde{g} \textit{ as above, and compute } \psi_{\tilde{g}} = \mathcal{F}^{-1}(\tilde{g}\widehat{\psi}).\end{aligned} \tag{18}$$

In the description above, the set E could clearly be replaced with $[-2\pi,-\pi) \cup [\pi,2\pi)$, or with any other "dyadic" set $[-2a,a) \cup [a,2a)$ for some $a > 0$.

2.5.3. *Wavelets of computationally elementary form*

We now give an account of s-elementary and MSF-wavelets. The two most elementary dyadic orthonormal wavelets are the *Haar wavelet* and *Shannon's wavelet* (also called the Littlewood–Paley wavelet).

The Haar wavelet is the function

$$\psi_H(t) = \begin{cases} 1, 0 \le t < \frac{1}{2} \\ -1, \frac{1}{2} \le t \le 1 \\ 0, \text{otherwise.} \end{cases} \tag{19}$$

In this case it is very easy to see that the dilates/translates

$$\{2^{\frac{n}{2}}\psi_H(2^n - \ell)\colon\ n, \ell \in \mathbb{Z}\}$$

are orthonormal, and an elementary argument shows that their span is dense in $L^2(\mathbb{R})$.

Shannon's wavelet is the $L^2(\mathbb{R})$-function with Fourier transform $\widehat{\psi}_S = \frac{1}{\sqrt{2\pi}}\chi_{E_0}$ where

$$E_0 = [-2\pi, -\pi) \cup [\pi, 2\pi). \tag{20}$$

The argument that $\widehat{\psi}_S$ is a wavelet is in a way even more transparent than for the Haar wavelet. And it has the advantage of generalizing nicely. For a simple argument, start from the fact that the exponents

$$\{e^{i\ell s}\colon\ n \in \mathbb{Z}\}$$

restricted to $[0, 2\pi]$ and normalized by $\frac{1}{\sqrt{2\pi}}$ is an orthonormal basis for $L^2[0, 2\pi]$. Write $E_0 = E_- \cup E_+$ where $E_- = [-2\pi, -\pi)$, $E_+ = [\pi, 2\pi)$. Since $\{E_- + 2\pi, E_+\}$ is a partition of $[0, 2\pi)$ and since the exponentials $e^{i\ell s}$ are invariant under translation by 2π, it follows that

$$\left\{ \frac{e^{i\ell s}}{\sqrt{2\pi}}\bigg|_{E_0}\colon\ n \in \mathbb{Z} \right\} \tag{21}$$

is an orthonormal basis for $L^2(E_0)$. Since $\widehat{T} = M_{e^{-is}}$, this set can be written

$$\{\widehat{T}^\ell \widehat{\psi}_s\colon\ \ell \in \mathbb{Z}\}. \tag{22}$$

Next, note that any "dyadic interval" of the form $J = [b, 2b)$, for some $b > 0$ has the property that $\{2^n J\colon\ n \in \mathbb{Z}\}$, is a partition of $(0, \infty)$. Similarly, any set of the form

$$\mathcal{K} = [-2a, -a) \cup [b, 2b) \tag{23}$$

for $a, b > 0$, has the property that

$$\{2^n \mathcal{K}\colon\ n \in \mathbb{Z}\}$$

is a partition of $\mathbb{R}\backslash\{0\}$. It follows that the space $L^2(\mathcal{K})$, considered as a subspace of $L^2(\mathbb{R})$, is a complete wandering subspace for the dilation unitary $(Df)(s) = \sqrt{2}\ f(2s)$. For each $n \in \mathbb{Z}$,

$$D^n(L^2(\mathcal{K})) = L^2(2^{-n}\mathcal{K}). \tag{24}$$

So $\bigoplus_n D^n(L^2(\mathcal{K}))$ is a direct sum decomposition of $L^2(\mathbb{R})$. In particular E_0 has this property. So

$$D^n \left\{ \frac{e^{i\ell s}}{\sqrt{2\pi}}\bigg|_{E_0}\colon\ \ell \in \mathbb{Z} \right\} = \left\{ \frac{e^{2^n i\ell s}}{\sqrt{2\pi}}\bigg|_{2^{-n}E_0}\colon\ \ell \in \mathbb{Z} \right\} \tag{25}$$

is an orthonormal basis for $L^2(2^{-n}E_0)$ for each n. It follows that

$$\{D^n \widehat{T}^\ell \widehat{\psi}_s\colon\ n, \ell \in \mathbb{Z}\}$$

is an orthonormal basis for $L^2(\mathbb{R})$. Hence $\{D^n T^\ell \psi_s\colon\ n, \ell \in \mathbb{Z}\}$ is an orthonormal basis for $L^2(\mathbb{R})$, as required.

The Haar wavelet can be generalized, and in fact Daubechies' well-known continuous compactly-supported wavelet is a generalization of the Haar wavelet. However, known generalization of the Haar wavelet are all more complicated and difficult to work with in hand-computations.

For our work, in order to proceed with developing an operator algebraic theory that had a chance of directly impacting concrete function-theoretic wavelet theory we needed a large supply of examples of wavelets which were elementary enough to work with. First, we found another "Shannon-type" wavelet in the literature. This was the Journe wavelet, which we found described on p. 136 in Daubechies book [14]. Its Fourier transform is $\widehat{\psi}_J = \frac{1}{\sqrt{2\pi}}\chi_{E_J}$, where

$$E_J = \left[-\frac{32\pi}{7}, -4\pi\right) \cup \left[-\pi, -\frac{4\pi}{7}\right) \cup \left[\frac{4\pi}{7}, \pi\right) \cup \left[4\pi, \frac{32\pi}{7}\right).$$

Then, thinking the old adage "where there's smoke there's fire!", we painstakingly worked out many more examples. So far, these are the basic building blocks in the *concrete* part of our theory. By this we mean the part of our theory that has had some type of direct impact on function-theoretic wavelet theory.

2.5.4. *Definition of wavelet set*

We define a *wavelet set* to be a measurable subset E of $\mathbb{R}$ for which $\frac{1}{\sqrt{2\pi}}\chi_E$ is the Fourier transform of a wavelet. The wavelet $\widehat{\psi}_E := \frac{1}{\sqrt{2\pi}}\chi_E$ is called *s-elementary* in [11].

It turns out that this class of wavelets was also discovered and systematically explored completely independently, and in about the same time period, by Guido Weiss (Washington University), his colleague and former student E. Hernandez (U. Madrid), and his students X. Fang and X. Wang. In [17,37, 38] they are called MSF (minimally supported frequency) wavelets. In signal processing, the parameter s, which is the independent variable for $\widehat{\psi}$, is the *frequency* variable, and the variable t, which is the independent variable for ψ, is the *time* variable. No function with support a subset of a wavelet set E of strictly smaller measure can be the Fourier transform of a wavelet.

Problem. Must the support of the Fourier transform of a wavelet contain a wavelet set? This question is open for dimension 1. It makes sense for any finite dimension.

2.5.5. *The spectral set condition*

From the argument above describing why Shannon's wavelet is, indeed, a wavelet, it is clear that *sufficient* conditions for E to be a wavelet set are

(i) the normalized exponential $\frac{1}{\sqrt{2\pi}}e^{i\ell s}$, $\ell \in \mathbb{Z}$, when restricted to E should constitute an orthonormal basis for $L^2(E)$ (in other words E is a *spectral set* for the integer lattice $\mathbb{Z}$),

and

(ii) the family $\{2^n E\colon\ n \in \mathbb{Z}\}$ of dilates of E by integral powers of 2 should constitute a measurable partition (i.e. a partition modulo null sets) of $\mathbb{R}$.

These conditions are also necessary. In fact if a set E satisfies (i), then for it to be a wavelet set it is obvious that (ii) must be satisfied. To show that (i) must be satisfied by a wavelet set E, consider the vectors

$$\widehat{D}^n \widehat{\psi}_E = \frac{1}{\sqrt{2\pi}} \chi_{2^{-n}E}, \qquad n \in \mathbb{Z}.$$

Since $\widehat{\psi}_E$ is a wavelet these must be orthogonal, and so the sets $\{2^n E\colon\ n \in\ \mathbb{Z}\}$ must be disjoint modulo null sets. It follows that $\{\frac{1}{\sqrt{2\pi}}e^{i\ell s}|_E\colon\ \ell \in \mathbb{Z}\}$ is not only an orthonormal set of vectors in $L^2(E)$, it must also *span* $L^2(E)$.

It is known from the theory of *spectral sets* (as an elementary special case) that a measurable set E satisfies (i) if and only if it is a generator of a measurable partition of $\mathbb{R}$ under translation by 2π (i.e. iff $\{E+2\pi n\colon\ n \in \mathbb{Z}\}$ is a measurable partition of $\mathbb{R}$). This result generalizes to spectral sets for the integral lattice in $\mathbb{R}^n$. For this elementary special case a direct proof is not hard.

2.5.6. *Translation and dilation congruence*

We say that measurable sets E, F are *translation congruent modulo* 2π if there is a measurable bijection $\phi\colon\ E \to F$ such that $\phi(s) - s$ is an integral multiple of 2π for each $s \in E$; or equivalently, if there is a measurable partition $\{E_n\colon\ n \in \mathbb{Z}\}$ of E such that

$$\{E_n + 2n\pi\colon\ n \in \mathbb{Z}\} \tag{26}$$

is a measurable partition of F. Analogously, define measurable sets G and H to be *dilation congruent modulo* 2 if there is a measurable bijection $\tau\colon\ G \to H$ such that for each $s \in G$ there is an integer n, depending on s, such that $\tau(s) = 2^n s$; or equivalently, if there is a measurable partition $\{G_n\}_{-\infty}^{\infty}$ of G such that

$$\{2^n G\}_{-\infty}^{\infty} \tag{27}$$

is a measurable partition of H. (Translation and dilation congruency modulo other positive numbers of course make sense as well.)

The following lemma is useful.

Lemma 7: *Let $f \in L^2(\mathbb{R})$, and let $E = supp(f)$. Then f has the property that*

$$\{e^{ins} f\colon\ n \in \mathbb{Z}\}$$

is an orthonormal basis for $L^2(E)$ if and only if

(i) E is congruent to $[0, 2\pi)$ modulo 2π, and
(ii) $|f(s)| = \frac{1}{\sqrt{2\pi}}$ a.e. on E.

If E is a measurable set which is 2π-translation congruent to $[0, 2\pi)$, then since

$$\left\{ \frac{e^{i\ell s}}{\sqrt{2\pi}}\bigg|_{[0,2\pi)} :\ \ell \in \mathbb{Z} \right\}$$

is an orthonormal basis for $L^2[0, 2\pi]$ and the exponentials $e^{i\ell s}$ are 2π-invariant, as in the case of Shannon's wavelet it follows that

$$\left\{ \frac{e^{i\ell s}}{\sqrt{2\pi}}\bigg|_{E} :\ \ell \in \mathbb{Z} \right\}$$

is an orthonormal basis for $L^2(E)$. Also, if E is 2π-translation congruent to $[0, 2\pi)$, then since

$$\{[0, 2\pi) + 2\pi n\colon\ n \in \mathbb{Z}\}$$

is a measurable partition of $\mathbb{R}$, so is

$$\{E + 2\pi n\colon\ n \in \mathbb{Z}\}.$$

These arguments can be reversed.

We say that a measurable subset $G \subset \mathbb{R}$ is a 2-*dilation generator* of a *partition* of $\mathbb{R}$ if the sets

$$2^n G := \{2^n s\colon\ s \in G\}, \qquad n \in \mathbb{Z} \tag{28}$$

are disjoint and $\mathbb{R}\setminus \cup_n 2^n G$ is a null set. Also, we say that $E \subset \mathbb{R}$ is a 2π-*translation generator of a partition* of $\mathbb{R}$ if the sets

$$E + 2n\pi := \{s + 2n\pi: \ s \in E\}, \qquad n \in \mathbb{Z}, \tag{29}$$

are disjoint and $\mathbb{R}\setminus \cup_n (E + 2n\pi)$ is a null set.

Lemma 8: *A measurable set $E \subseteq \mathbb{R}$ is a 2π-translation generator of a partition of $\mathbb{R}$ if and only if, modulo a null set, E is translation congruent to $[0, 2\pi)$ modulo 2π. Also, a measurable set $G \subseteq \mathbb{R}$ is a 2-dilation generator of a partition of $\mathbb{R}$ if and only if, modulo a null set, G is a dilation congruent modulo 2 to the set $[-2\pi, -\pi) \cup [\pi, 2\pi)$.*

2.5.7. *A criterion*

The following is a useful criterion for wavelet sets. It was published independently by Dai–Larson in [11] and by Fang–Wang in [17] at about the same time in December, 1994. In fact, it is amusing that the two papers had been submitted within two days of each other; only much later did we even learn of each others work and of this incredible timing.

Proposition 9: *Let $E \subseteq \mathbb{R}$ be a measurable set. Then E is a wavelet set if and only if E is both a 2-dilation generator of a partition (modulo null sets) of $\mathbb{R}$ and a 2π-translation generator of a partition (modulo null sets) of $\mathbb{R}$. Equivalently, E is a wavelet set if and only if E is both translation congruent to $[0, 2\pi)$ modulo 2π and dilation congruent to $[-2\pi, -\pi) \cup [\pi, 2\pi)$ modulo 2.*

Note that a set is 2π-translation congruent to $[0, 2\pi)$ iff it is 2π-translation congruent to $[-2\pi, -\pi) \cup [\pi, 2\pi)$. So the last sentence of Proposition 9 can be stated: A measurable set E is a wavelet set if and only if it is both 2π-translation and 2-dilation congruent to the Littlewood–Paley set $[-2\pi, -\pi) \cup [\pi, 2\pi)$.

2.6. *Phases*

If E is a wavelet set, and if $f(s)$ is any function with support E which has constant modulus $\frac{1}{\sqrt{2\pi}}$ on E, then $\mathcal{F}^{-1}(f)$ is a wavelet. Indeed, by Lemma 7 $\{\widehat{T}^\ell f: \ \in \mathbb{Z}\}$ is an orthonormal basis for $L^2(E)$, and since the sets $2^n E$ partition $\mathbb{R}$, so $L^2(E)$ is a complete wandering subspace for $\widehat{D}$, it follows that $\{\widehat{D}^n \widehat{T}^\ell f: \ n, \ell \in \mathbb{Z}\}$ must be an orthonormal basis for $L^2(\mathbb{R})$, as required. In [17, 37, 38] the term MSF-wavelet includes this type of wavelet.

So MSF-wavelets can have arbitrary phase and s-elementary wavelets have phase 0. *Every* phase is attainable in the sense of chapter 3 for an MSF or s-elementary wavelet.

2.6.1. *Some examples of one-dimensional wavelet sets*

It is usually easy to determine, using the dilation-translation criteria, in Proposition 9, whether a given finite union of intervals is a wavelet set. In fact, to verify that a given "candidate" set E is a wavelet set, it is clear from the above discussion and criteria that it suffices to do two things.

(1) Show, by appropriate partitioning, that E is 2-dilation-congruent to a set of the form $[-2a, -a) \cup [b, 2b)$ for some $a, b > 0$.

and

(2) Show, by appropriate partitioning, that E is 2π-translation-congruent to a set of the form $[c, c + 2\pi)$ for some real number c.

On the other hand, wavelet sets suitable for testing hypotheses, can be quite difficult to construct. There are very few "recipes" for wavelet sets, as it were. Many families of such sets have been constructed for reasons including perspective, experimentation, testing hypotheses, etc., including perhaps the pure enjoyment of doing the computations – which are somewhat "puzzle-like" in nature. In working with the theory it is nice (and in fact necessary) to have a large supply of wavelets on hand that permit relatively simple analysis.

For this reason we take the opportunity here to present for the reader a collection of such sets, mainly taken from [11], leaving most of the "fun" in verifying that they are indeed wavelet sets to the reader.

We refer the reader to [12] for a proof of the existence of wavelet sets in $\mathbb{R}^{(n)}$, and a proof that there are sufficiently many to generate the Borel structure of $\mathbb{R}^{(n)}$. These results are true for arbitrary expansive dilation factors. Some concrete examples in the plane were subsequently obtained by Soardi and Weiland, and others were obtained by Gu and Speegle. Two had also been obtained by Dai for inclusion in the revised concluding remarks section of our Memoir [11].

In these examples we will usually write intervals as half-open intervals $[\cdot, \ )$ because it is easier to verify the translation and dilation congruency relations (1) and (2) above when wavelet sets are written thus, even though in actuality the relations need only hold modulo null sets.

(i) As mentioned above, an example due to Journe of a wavelet which admits no multiresolution analysis is the s-elementary wavelet with wavelet set

$$\left[-\frac{32\pi}{7}, -4\pi\right) \cup \left[-\pi, -\frac{4\pi}{7}\right) \cup \left[\frac{4\pi}{7}, \pi\right) \cup \left[4\pi, \frac{32\pi}{7}\right).$$

To see that this satisfies the criteria, label these intervals, in order, as J_1, J_2, J_3, J_4 and write $J = \cup J_i$. Then

$$J_1 \cup 4J_2 \cup 4J_3 \cup J_4 = \left[-\frac{32\pi}{7}, -\frac{16\pi}{7}\right) \cup \left[\frac{16\pi}{7}, \frac{32\pi}{7}\right).$$

This has the form $[-2a, a) \cup [b, 2b)$ so is a 2-dilation generator of a partition of $\mathbb{R}\backslash\{0\}$. Then also observe that

$$\{J_1 + 6\pi, J_2 + 2\pi, J_3, J_4 - 4\pi\}$$

is a partition of $[0, 2\pi)$.

(ii) The Shannon (or Littlewood–Paley) set can be generalized. For any $-\pi < \alpha < \pi$, the set

$$E_\alpha = [-2\pi + 2\alpha, -\pi + \alpha) \cup [\pi + \alpha, 2\pi + 2\alpha)$$

is a wavelet set. Indeed, it is clearly a 2-dilation generator of a partition of $\mathbb{R}\backslash\{0\}$, and to see that it satisfies the translation congruency criterion for $-\pi < \alpha \leq 0$ (the case $0 < \alpha < \pi$ is analogous) just observe that

$$\{[-2\pi + 2\alpha, 2\pi) + 4\pi, [-2\pi, -\pi + \alpha) + 2\pi, [\pi + \alpha, 2\pi + 2\alpha)\}$$

is a partition of $[0, 2\pi)$. It is clear that ψ_{E_α} is then a continuous (in $L^2(\mathbb{R})$-norm) path of s-elementary wavelets. Note that

$$\lim_{\alpha \to \pi} \widehat{\psi}_{E_\alpha} = \frac{1}{\sqrt{2\pi}} \chi_{[2\pi, 4\pi)}.$$

This is *not* the Fourier transform of a wavelet because the set $[2\pi, 4\pi)$ is not a 2-dilation generator of a partition of $\mathbb{R}\backslash\{0\}$. So

$$\lim_{\alpha \to \pi} \psi_{E_\alpha}$$

is not an orthogonal wavelet. (It is what is known as a Hardy wavelet because it generates an orthonormal basis for $H^2(\mathbb{R})$ under dilation and translation.) This example demonstrates that $\mathcal{W}(D, T)$ is *not* closed in $L^2(\mathbb{R})$.

(iii) Journe's example above can be extended to a path. For $-\frac{\pi}{7} \leq \beta \leq \frac{\pi}{7}$ the set

$$J_\beta = \left[-\frac{32\pi}{7}, -4\pi + 4\beta\right) \cup \left[-\pi + \beta, -\frac{4\pi}{7}\right) \cup \left[\frac{4\pi}{7}, \pi + \beta\right) \cup \left[4\pi + 4\beta, 4\pi + \frac{4\pi}{7}\right)$$

is a wavelet set. The same argument in (i) establishes dilation congruency. For translation, the argument in (i) shows congruency to $[4\beta, 2\pi + 4\beta)$ which is in turn congruent to $[0, 2\pi)$ as required. Observe that here, as opposed to in (ii) above, the limit of ψ_{J_β} as β approaches the boundary point $\frac{\pi}{7}$ *is* a wavelet. Its wavelet set is a union of 3 disjoint intervals.

(iv) Let $A \subseteq [\pi, \frac{3\pi}{2})$ be an arbitrary measurable subset. Then there is a wavelet set W, such that $W \cap [\pi, \frac{3\pi}{2}) = A$. For the construction, let

$$\begin{aligned} B &= [2\pi, 3\pi) \backslash 2A, \\ C &= \left[-\pi, -\frac{\pi}{2}\right) \backslash (A - 2\pi) \\ \text{and} \quad D &= 2A - 4\pi. \end{aligned}$$

Let

$$W = \left[\frac{3\pi}{2}, 2\pi\right) \cup A \cup B \cup C \cup D.$$

We have $W \cap [\pi, \frac{3\pi}{2}) = A$. Observe that the sets $[\frac{3\pi}{2}, 2\pi)$, A, B, C, D, are disjoint. Also observe that the sets

$$\left[\frac{3\pi}{2}, 2\pi\right), A, \frac{1}{2}B, 2C, D,$$

are disjoint and have union $[-2\pi, -\pi) \cup [\pi, 2\pi)$. In addition, observe that the sets

$$\left[\frac{3\pi}{2}, 2\pi\right), A, B - 2\pi, C + 2\pi, D + 2\pi,$$

are disjoint and have union $[0, 2\pi)$. Hence W is a wavelet set.

(v) Wavelet sets for arbitrary (not necessarily integral) dilation factors other then 2 exist. For instance, if $d \geq 2$ is arbitrary, let

$$\begin{aligned} A &= \left[-\frac{2d\pi}{d+1}, -\frac{2\pi}{d+1}\right), \\ B &= \left[\frac{2\pi}{d^2 - 1}, \frac{2\pi}{d+1}\right), \\ C &= \left[\frac{2d\pi}{d+1}, \frac{2d^2\pi}{d^2 - 1}\right) \end{aligned}$$

and let $G = A \cup B \cup C$. Then G is d-wavelet set. To see this, note that $\{A + 2\pi, B, C\}$ is a partition of an interval of length 2π. So G is 2π-translation-congruent to $[0, 2\pi)$. Also, $\{A, B, d^{-1}C\}$ is a partition of the set $[-d\alpha, -\alpha) \cup [\beta, d\beta)$ for $\alpha = \frac{2\pi}{d^2-1}$, and $\beta = \frac{2\pi}{d^2-1}$, so from this form it follows that $\{d^n G\colon\ n \in \mathbb{Z}\}$ is a partition of $\mathbb{R}\backslash\{0\}$. Hence if $\psi := \mathcal{F}^{-1}(\frac{1}{\sqrt{2\pi}}\chi_G)$, it follows that $\{d^{\frac{n}{2}}\psi(d^n t - \ell)\colon\ n, \ell \in \mathbb{Z}\}$ is orthonormal basis for $L^2(\mathbb{R})$, as required.

2.7. *Operator-theoretic interpolation of wavelets: The special case of wavelet sets*

Let E, F be a pair of wavelet sets. Then for (a.e.) $x \in E$ there is a unique $y \in F$ such that $x - y \in 2\pi\mathbb{Z}$. This is the *translation congruence* property of wavelet sets. Also, for (a.e.) $x \in E$ there is a unique $z \in F$ such that $\frac{x}{z}$ is an integral power of 2. This is the *dilation congruence* property of wavelet sets. (See section 2.5.6.)

There is a natural *closed-form algorithm* for the *interpolation unitary* $V_{\psi_E}^{\psi_F}$ which maps the wavelet basis for $\widehat{\psi}_E$ to the wavelet basis for $\widehat{\psi}_F$. Indeed, using both the translation and dilation congruence properties of $\{E, F\}$, one can explicitly compute a (unique) measure-preserving transformation $\sigma := \sigma_E^F$ mapping $\mathbb{R}$ onto $\mathbb{R}$ which has the property that $V_{\psi_E}^{\psi_F}$ is identical with the *composition operator* defined by:

$$f \mapsto f \circ \sigma^{-1}$$

for all $f \in L^2(\mathbb{R})$. With this formulation, compositions of the maps σ between different pairs of wavelet sets are not difficult to compute, and thus products of the corresponding interpolation unitaries can be computed in terms of them.

2.7.1. *The interpolation map σ*

Let E and F be arbitrary wavelet sets. Let $\sigma\colon\ E \to F$ be the 1-1, onto map implementing the 2π-translation congruence. Since E and F both generated partitions of $\mathbb{R}\backslash\{0\}$ under dilation by powers of 2, we may extend σ to a 1-1 map of $\mathbb{R}$ onto $\mathbb{R}$ by defining $\sigma(0) = 0$, and

$$\sigma(s) = 2^n \sigma(2^{-n} s) \quad \text{for} \quad s \in 2^n E, \quad n \in \mathbb{Z}. \tag{30}$$

We adopt the notation σ_E^F for this, and call it the *interpolation map* for the ordered pair (E, F).

Lemma 10: *In the above notation, σ_E^F is a measure-preserving transformation from $\mathbb{R}$ onto $\mathbb{R}$.*

Proof: Let $\sigma := \sigma_E^F$. Let $\Omega \subseteq \mathbb{R}$ be a measurable set. Let $\Omega_n = \Omega \cap 2^n E$, $n \in \mathbb{Z}$, and let $E_n = 2^{-n}\Omega_n \subseteq E$. Then $\{\Omega_n\}$ is a partition of Ω, and we have $m(\sigma(E_n)) = m(E_n)$ because the restriction of σ to E is measure-preserving. So

$$\begin{aligned} m(\sigma(\Omega)) &= \sum_n m(\sigma(\Omega_n)) = \sum_n m(2^n\sigma(E_n)) \\ &= \sum_n 2^n m(\sigma(E_n)) = \sum_n 2^n m(E_n) \\ &= \sum_n m(2^n E_n) = \sum_n m(\Omega_n) = m(\Omega). \end{aligned}$$

□

A function $f\colon \mathbb{R} \to \mathbb{R}$ is called *2-homogeneous* if $f(2s) = 2f(s)$ for all $s \in \mathbb{R}$. Equivalently, f is 2-homogeneous iff $f(2^n s) = 2^n f(s)$, $s \in \mathbb{R}$, $n \in \mathbb{Z}$. Such a function is completely determined by its values on any subset of $\mathbb{R}$ which generates a partition of $\mathbb{R}\backslash\{0\}$ by 2-dilation. So σ_E^F is the (unique) 2-homogeneous extension of the 2π-transition congruence $E \to F$. The set of all 2-homogeneous measure-preserving transformations of $\mathbb{R}$ clearly forms a group under composition. Also, the composition of a 2-dilation-periodic function f with a 2-homogeneous function g is (in either order) 2-dilation periodic. We have $f(g(2s)) = f(2g(s)) = f(g(s))$ and $g(f(2s)) = g(f(s))$. These facts will be useful.

2.7.2. *An algorithm for the interpolation unitary*

Now let

$$U_E^F := U_{\sigma_E^F}, \tag{31}$$

where if σ is any measure-preserving transformation of $\mathbb{R}$ then U_σ denotes the composition operator defined by $U_\sigma f = f \circ \sigma^{-1}$, $f \in L^2(\mathbb{R})$. Clearly $(\sigma_E^F)^{-1} = \sigma_F^E$ and $(U_E^F)^* = U_F^E$. We have $U_E^F\widehat{\psi}_E = \widehat{\psi}_F$ since $\sigma_E^F(E) = F$. That is,

$$U_E^F\widehat{\psi}_E = \widehat{\psi}_E \circ \sigma_F^E = \frac{1}{\sqrt{2\pi}}\chi_E \circ \sigma_F^E = \frac{1}{\sqrt{2\pi}}\chi_F = \widehat{\psi}_F.$$

Proposition 11: *Let E and F be arbitrary wavelet sets. Then $U_E^F \in \mathcal{C}_{\widehat{\psi}_E}(\widehat{D},\widehat{T})$. Hence $\mathcal{F}^{-1}U_E^F\mathcal{F}$ is the interpolation unitary for the ordered pair (ψ_E, ψ_F).*

Proof: Write $\sigma = \sigma_E^F$ and $U_\sigma = U_E^F$. We have $U_\sigma \widehat{\psi}_E = \widehat{\psi}_F$ since $\sigma(E) = F$. We must show

$$U_\sigma \widehat{D}^n \widehat{T}^l \widehat{\psi}_E = \widehat{D}^n \widehat{T}^l U_\sigma \widehat{\psi}_E, \quad n, l \in \mathbb{Z}.$$

We have

$$\begin{aligned}(U_\sigma \widehat{D}^n \widehat{T}^l \widehat{\psi}_E)(s) &= (U_\sigma \widehat{D}^n e^{-ils} \widehat{\psi}_E)(s)\\ &= U_\sigma 2^{-\frac{n}{2}} e^{-il2^{-n}s} \widehat{\psi}_E(2^{-n}s)\\ &= 2^{-\frac{n}{2}} e^{-il2^{-n}\sigma^{-1}(s)} \widehat{\psi}_E(2^{-n}\sigma^{-1}(s))\\ &= 2^{-\frac{n}{2}} e^{-il\sigma^{-1}(2^{-n}s)} \widehat{\psi}_E(\sigma^{-1}(2^{-n}s))\\ &= 2^{-\frac{n}{2}} e^{-il\sigma^{-1}(2^{-n}s)} \widehat{\psi}(2^{-n}s).\end{aligned}$$

This last term is nonzero iff $2^{-n}s \in F$, in which case $\sigma^{-1}(2^{-n}s) = \sigma_F^E(2^{-n}s)$ $= 2^{-n}s + 2\pi k$ for some $k \in \mathbb{Z}$ since σ_F^E is a 2π-translation-congruence on F. It follows that $e^{-il\sigma^{-2}(2^{-n}s)} = e^{-il2^{-n}s}$. Hence we have

$$\begin{aligned}(U_\sigma \widehat{D}^n \widehat{T}^l \widehat{\psi}_E)(s) &= 2^{-\frac{n}{2}} e^{-ils^{-2n}s} \widehat{\psi}_F(2^{-n}s)\\ &= (\widehat{D}^n \widehat{T}^l \widehat{\psi}_F)(s)\\ &= (\widehat{D}^n \widehat{T}^l U_\sigma \widehat{\psi}_E)(s).\end{aligned}$$

We have shown $U_E^F \in \mathcal{C}_{\widehat{\psi}_E}(\widehat{D}, \widehat{T})$. Since $U_E^F \widehat{\psi}_E = \widehat{\psi}_F$, the uniqueness part of Proposition 1 shows that $\mathcal{F}^{-1} U_E^F \mathcal{F}$ must be the interpolation unitary for (ψ_E, ψ_F). □

2.8. *The interpolation unitary normalizes the commutant*

Proposition 12: *Let E and F be arbitrary wavelet sets. Then the interpolation unitary for the ordered pair (ψ_E, ψ_F) normalizes $\{D, T\}'$.*

Proof: By Proposition 11 we may work with U_E^F in the Fourier transform domain. By Theorem 6, the generic element of $\{\widehat{D}, \widehat{T}\}'$ has the form M_h for some 2-dilation-periodic function $h \in L^\infty(\mathbb{R})$. Write $\sigma = \sigma_E^F$ and $U_\sigma = U_E^F$. Then

$$U_\sigma^{-1} M_h U_\sigma = M_{h \circ \sigma^{-1}}. \tag{32}$$

So since the composition of a 2-dilation-periodic function with a 2-homogeneous function is 2-dilation-periodic, the proof is complete. □

2.8.1. $\mathcal{C}_\psi(D,T)$ *is nonabelian*

It can also be shown ([11, Theorem 5.2 (iii)]) that if E, F are wavelet sets with $E \neq F$ then U_E^F is not contained in the double commutant $\{\widehat{D},\widehat{T}\}''$. So since U_E^F and $\{\widehat{D},\widehat{T}\}'$ are both contained in the local commutant of $\mathcal{U}_{\widehat{D},\widehat{T}}$ at $\widehat{\psi}_E$, this proves that $\mathcal{C}_{\widehat{\psi}_E}(\widehat{D},\widehat{T})$ is nonabelian. In fact (see [11, Proposition 1.8]) this can be used to show that $\mathcal{C}_\psi(D,T)$ is nonabelian for every wavelet ψ. We suspected this, but we could not prove it until we discovered the "right" way of doing the needed computation using s-elementary wavelets.

The above shows that a pair (E,F) of wavelets sets (or, rather, their corresponding s-elementary wavelets) admits operator-theoretic interpolation if and only if Group$\{U_E^F\}$ is contained in the local commutant $\mathcal{C}_{\widehat{\psi}_E}(\widehat{D},\widehat{T})$, since the requirement that U_E^F normalizes $\{\widehat{D},\widehat{T}\}'$ is automatically satisfied. It is easy to see that this is equivalent to the condition that for each $n \in \mathbb{Z}$, σ^n is a 2π-congruence of E in the sense that $(\sigma^n(s)-s)/2\pi \in \mathbb{Z}$ for all $s \in E$, which in turn implies that $\sigma^n(E)$ is a wavelet set for all n. Here $\sigma = \sigma_E^F$. This property holds trivially if σ is *involutive* (i.e. $\sigma^2 =$ identity).

2.8.2. *The coefficient criterion*

In cases where "torsion" is present, so $(\sigma_E^F)^k$ is the identity map for some finite integer k, the von Neumann algebra generated by $\{\widehat{D},\widehat{T}\}'$ and $U := U_E^F$ has the simple form

$$\left\{\sum_{n=0}^{k} M_{h_n}U^n : \; h_n \in L^\infty(\mathbb{R}) \text{ with } h_n(2s) = h_n(s), \quad s \in \mathbb{R}\right\},$$

and so each member of this "interpolated" family of wavelets has the form

$$\frac{1}{\sqrt{2\pi}}\sum_{n=0}^{k} h_n(s)\chi_{\sigma^n(E)} \tag{33}$$

for 2-dilation periodic "coefficient" functions $\{h_n(s)\}$ which satisfy the necessary and sufficient condition that the operator

$$\sum_{n=0}^{k} M_{h_n}U^n \tag{34}$$

is unitary.

A standard computation shows that the map θ sending $\sum_0^k M_{h_n}U^n$ to the $k\times k$ function matrix (h_{ij}) given by

$$h_{ij} = h_{\alpha(i,j)} \circ \sigma^{-i+1} \tag{35}$$

where $\alpha(i,j) = (i+1)$ modulo k, is a $*$-isomorphism. This matricial algebra is the cross-product of $\{D,T\}'$ by the $*$-automorphism $ad(U_E^F)$ corresponding to conjugation with U_E^F. For instance, if $k=3$ then θ maps

$$M_{h_1} + M_{h_2}U_E^F + M_{h_3}(U_E^F)^2$$

to

$$\begin{pmatrix} h_1 & h_2 & h_3 \\ h_3 \circ \sigma^{-1} & h_1 \circ \sigma^{-1} & h_2 \circ \sigma^{-1} \\ h_2 \circ \sigma^{-2} & h_3 \circ \sigma^{-2} & h_1 \circ \sigma^{-2} \end{pmatrix}. \tag{36}$$

This shows that $\sum_0^k M_{h_n}U^n$ is a unitary operator iff the scalar matrix $(h_{ij})(s)$ is unitary for almost all $s \in \mathbb{R}$. Unitarity of this matrix-valued function is called the *Coefficient Criterion* in [11], and the functions h_i are called the interpolation coefficients. This leads to formulas for families of wavelets which are new to wavelet theory.

2.9. *Interpolation pairs of wavelet sets*

For many interesting cases of note, the interpolation map σ_E^F will in fact be an *involution* of $\mathbb{R}$ (i.e. $\sigma \circ \sigma = id$, where $\sigma := \sigma_E^F$, and where id denotes the identity map). So torsion *will* be present, as in the above section, and it will be present in an essentially simple form. The corresponding interpolation unitary will be a *symmetry* in this case (i.e. a selfadjoint unitary operator with square I).

It is curious to note that verifying a simple operator equation $U^2 = I$ directly by matricial computation can be extremely difficult. It is much more computationally feasible to verify an equation such as this by pointwise (a.e.) verifying explicitly the relation $\sigma \circ \sigma = id$ for the interpolation map. In [11] we gave a number of examples of interpolation pairs of wavelet sets. We give below a collection of examples that has not been previously published: Every pair sets from the Journe family is an interpolation pair.

2.10. *Journe family interpolation pairs*

Consider the parameterized path of *generalized Journe* wavelet sets given in [11, Example 4.5(iii)]. We have

$$J_\beta = \left[-\frac{32\pi}{7}, -4\pi - 4\beta\right) \cup \left[-\pi - \beta, -\frac{4\pi}{7}\right) \cup \left[\frac{4\pi}{7}, \pi + \beta\right) \cup \left[4\pi + 4\beta, \frac{32\pi}{7}\right)$$

where the set of parameters β ranges $-\frac{\pi}{7} \leq \beta \leq \frac{\pi}{7}$.

Proposition 13: *Every pair $(J_{\beta_1}, J_{\beta_2})$ is an interpolation pair.*

Proof: Let $\beta_1, \beta_2 \in \left[-\frac{\pi}{7}, \frac{\pi}{7}\right)$ with $\beta_1 < \beta_2$. Write $\sigma = \sigma_{J_{\beta_2}}^{J_{\beta_1}}$. We need to show that

$$\sigma^2(x) = x \tag{*}$$

for all $x \in \mathbb{R}$. Since σ is 2-homogeneous, it suffices to verify (*) only for $x \in J_{\beta_1}$. For $x \in J_{\beta_1} \cap J_{\beta_2}$ we have $\sigma(x) = x$, hence $\sigma^2(x) = x$. So we only need to check (*) for $x \in (J_{\beta_1} \backslash J_{\beta_2})$. We have

$$J_{\beta_1} \backslash J_{\beta_2} = [-\pi + \beta_1, -\pi + \beta_2) \cup [4\pi + 4\beta_1, 4\pi + 4\beta_2).$$

It is useful to also write

$$J_{\beta_2} \backslash J_{\beta_1} = [-4\pi + 4\beta_1, -4\pi + 4\beta_2) \cup [\pi + \beta_1, \pi + \beta_2).$$

On $[-\pi+\beta_1, -\pi+\beta_2)$ we have $\sigma(x) = x+2\pi$, which lies in $[\pi+\beta_1, \pi+\beta_2)$. If we multiply this by 4, we obtain $4\sigma(x) \in [4\pi + 4\beta_1, 4\pi + 4\beta_2) \subset J_{\beta_1}$. And on $[4\pi + 4\beta_1, 4\pi + 4\beta_2)$ we clearly have $\sigma(x) = x - 8\pi$, which lies in $[-4\pi + 4\beta_1, -4\pi + 4\beta_2)$.

So for $x \in [-\pi + \beta_1, -\pi + \beta_2)$ we have

$$\sigma^2(x) = \sigma(\sigma(x)) = \frac{1}{4}\sigma(4\sigma(x)) = \frac{1}{4}[4\sigma(x) - 8\pi] = \sigma(x) - 2\pi = x + 2\pi - 2\pi = x.$$

On $[4\pi + 4\beta_1, 4\pi + 4\beta_2)$ we have $\sigma(x) = x - 8\pi$, which lies in $[-4\pi + 4\beta_1, -4\pi + 4\beta_2)$. So $\frac{1}{4}\sigma(x) \in [-\pi + \beta_1, -\pi + \beta_2)$. Hence

$$\sigma\left(\frac{1}{4}\sigma(x)\right) = \frac{1}{4}\sigma(x) + 2\pi$$

and thus

$$\sigma^2(x) = 4\sigma\left(\frac{1}{4}\sigma(x)\right) = 4\left[\frac{1}{4}\sigma(x) + 2\pi\right] = \sigma(x) + 8\pi = x - 8\pi + 8\pi = x$$

as required.

We have shown that for all $x \in J_{\beta_1}$ we have $\sigma^2(x) = x$. This proves that $(J_{\beta_1}, J_{\beta_2})$ is an interpolation pair. □

3. Unitary Systems and Frames

In [33] we developed an operator-theoretic approach to discrete frame theory (i.e. frame sequences, as opposed to continuous frame transforms) on a separable Hilbert space. We then applied it to an investigation of frame vectors for unitary systems, frame wavelets and group representations. The starting-point idea, which is pretty simple-minded in fact, is to realize any frame sequence for a Hilbert space H as a compression of a Riesz basis for a larger Hilbert space. In other words, a frame is a sequence of vectors in a Hilbert space which *dilates*, (in the operator-theoretic or geometric sense, as opposed to the function-theoretic sense of multiplication of the independent variable of a function by a dilation constant), or *extends*, to a (Riesz) basis for a larger space. From this idea much can be developed, and some new perspective can be given to certain concepts that have been used in engineering circles for many years. See section 3.2. below.

3.1. *Basics on frames*

Let H be a separable complex Hilbert space. Let $B(H)$ denote the algebra of all bounded linear operators on H. Let $\mathbb{N}$ denote the natural numbers, and $\mathbb{Z}$ the integers. We will use $\mathbb{J}$ to denote a generic countable (or finite) index set such as $\mathbb{Z}, \mathbb{N}, \mathbb{Z}^{(2)}$, $\mathbb{N} \cup \mathbb{N}$ etc.

A sequence $\{x_j\colon\ j \in \mathbb{N}\}$ of vectors in H is called a *frame* if there are constants $A, B > 0$ such that

$$A\|x\|^2 \leq \sum_j |\langle x, x_j\rangle|^2 \leq B\|x\|^2$$

for all $x \in H$. The optimal constants (maximal for A and minimal for B) are called the *frame bounds.* The frame $\{x_j\}$ is called a *tight frame* if $A = B$, and is called *Parseval* if $A = B = 1$. (Originally, in [33] and in a number of subsequent papers, the term *normalized tight frame* was used for this. However, this term had also been applied by Benedetto and Fickus [5] for another concept: a tight frame of unit vectors; what we now call a uniform tight frame, or spherical frame. So, after all parties involved, the name *Parseval* was adopted. It makes a lot of sense, because a Parseval frame is precisely a frame which satisfies Parseval's identity.) A sequence $\{x_j\}$ is defined to be a *Riesz basis* if it is a frame and is also a basis for H in the sense that for each $x \in H$ there is a *unique* sequence $\{\alpha_j\}$ in $\mathbb{C}$ such that $x = \sum \alpha_j x_j$ with the convergence being in norm. We note that a Riesz basis is also defined to be a basis which is obtained from an orthonormal basis by applying a bounded linear invertible operator. This is equivalent to the first definition. It should be noted that in Hilbert spaces the Riesz bases are precisely the bounded unconditional bases. We will say that frames $\{x_j\colon\ j \in \mathbb{J}\}$ and $\{y_j\colon\ j \in \mathbb{J}\}$ on Hilbert spaces H, K, respectively, are *unitarily equivalent* if there is a unitary $U\colon\ H \to K$ such that $Ux_j = y_j$ for all $j \in \mathbb{J}$. We will say that they are *similar* (or *isomorphic*) if there is a bounded linear invertible operator $T\colon\ H \to K$ such that $Tx_j = y_j$ for all $j \in \mathbb{J}$.

Example 14: Let $K = L^2(\mathbb{T})$ where $\mathbb{T}$ is the unit circle and measure is normalized Lebesgue measure, and let $\{e^{ins}: n \in \mathbb{Z}\}$ be the standard orthonormal basis for $L^2(\mathbb{T})$. If $E \subseteq \mathbb{T}$ is any measurable subset then $\{e^{ins}|_E: n \in \mathbb{Z}\}$ is a Parseval frame for $L^2(E)$. This can be viewed as obtained from the single vector χ_E by applying all integral powers of the (unitary) multiplication operator $M_{e^{is}}$. It turns out that these are all (for different E) unitarily *inequivalent*. This is an example of a Parseval frame which is generated by the action of a unitary group on a single vector. This can be compared with the definition of a *frame wavelet*. (As one might expect, a single function ψ in $L^2(\mathbb{R})$ which generates a *frame* for $L^2(\mathbb{R})$ under the action of $\mathcal{U}_{D,T}$ is called a *frame wavelet*.)

3.2. *Dilation of frames: The discrete version of Naimark's theorem*

Now let $\{x_n\}_{n\in\mathbb{J}}$ be a Parseval frame and let $\theta\colon\ H \to K := l^2(\mathbb{J})$ be the usual *analysis operator* (this was called the *frame transform* in [33]) defined by $\theta(x) := (\langle x, x_n\rangle)_{n\in\mathbb{J}}$. This is obviously an isometry. Let P be the orthogonal projection from K onto $\theta(H)$. Denote the standard orthonormal basis for $l^2(\mathbb{J})$ by $\{e_j\colon\ j \in \mathbb{J}\}$. For any $m \in \mathbb{J}$, we have

$$\begin{aligned}\langle \theta(x_m), Pe_n\rangle &= \langle P\theta(x_m), e_n\rangle = \langle \theta(x_m), e_n\rangle \\ &= \langle x_m, x_n\rangle = \langle \theta(x_m), \theta(x_m)\rangle.\end{aligned}$$

It follows easily that $\theta(x_n) = Pe_n$, $n \in \mathbb{J}$. Identifying H with $\theta(H)$, this shows indeed that every Parseval frame can be realized by compressing an orthonormal basis, as claimed earlier.

This can actually be viewed as a special case (probably the simplest possible special case) of an old theorem of Naimark concerning operator algebras and dilation of positive operator valued measures to projection valued measures. The connection between Naimark's theorem and the dilation result for Parseval frames, and that the latter can be viewed as a special case of the former, was pointed out to me by Chandler Davis and Dick Kadison in a conference (COSY-1999: The Canadian Operator Algebra Symposium, Prince Edward Island, May 1999).

3.3. *Complements of frames*

It is useful to note that P will equal I iff $\{x_n\}$ is a basis. Indeed, if $P \neq I$, then choose $z \neq 0$, $z \in (I - P)K$, and write $z = \sum \alpha_n e_n$ for some sequence $\alpha_n \in \mathbb{C}$. Then $0 = Pz = \sum \alpha_n \theta(x_n)$, and not all the scalars α_n are zero.

Hence $\{x_n\}$ is not topologically linearly independent so cannot even be a Schauder basis. On the other hand if $P = I$ then $\{x_n\}$ is obviously an orthonormal basis.

Suppose $\{x_n\}_{n\in\mathbb{J}}$ is a Parseval frame for H, and let θ, P, K, e_n be as above. Let $M = (I - P)K$. Then $y_n := (I - P)e_n$ is a Parseval frame on M which is *complementary* to $\{x_n\}$ in the sense that the inner direct sum $\{x_n \oplus y_n\colon\ n \in \mathbb{J}\}$ is an orthonormal basis for the direct sum Hilbert space $H \oplus M$. Moreover there is uniqueness: The extension of a tight frame to an orthonormal basis described in the above paragraph is unique up to unitary equivalence. That is if N is another Hilbert space and $\{z_n\}$ is a tight frame for N such that $\{x_n \oplus z_n\colon\ n \in \mathbb{J}\}$ is an orthonormal basis for $H \oplus N$, then there is a unitary transformation U mapping M onto N such that $Uy_n = z_n$ for all n. In particular, $\dim M = \dim N$.

If $\{x_j\}$ is a Parseval frame, we will call any Parseval frame $\{z_j\}$ such that $\{x_j \oplus z_j\}$ is an orthonormal basis for the direct sum space, a *strong complement* to $\{x_j\}$. So every Parseval frame has a strong complement which is unique up to unitary equivalence. More generally, if $\{y_j\}$ is a general frame we will call any frame $\{w_j\}$ such that $\{y_j \oplus w_j\}$ is a Riesz basis for the direct sum space a *complementary* frame (or *complement*) to $\{x_j\}$.

The notion of strong complement has a natural generalization. Let $\{x_n\}_{n\in\mathbb{J}}$ and $\{y_n\}_{n\in\mathbb{J}}$ be Parseval frames in Hilbert spaces H, K, respectively, indexed by the same set $\mathbb{J}$. Call these two frames *strongly disjoint* if the (inner) direct sum $\{x_n \oplus y_n\colon\ n \in \mathbb{J}\}$ is a Parseval frame for the direct sum Hilbert space $H \oplus K$. It is not hard to see that this property of strong disjointness is equivalent to the property that the ranges of their analysis operators are orthogonal in $l^2(\mathbb{J})$. More generally, we call a k-tuple of Parseval frames $(\{z_{1n}\}_{n\in\mathbb{J}}, \dots, \{z_{kn}\}_{n\in\mathbb{J}})$ in Hilbert spaces $H_1, \dots, H_k$, respectively, a *strongly disjoint* k-tuple if $\{z_{1n} \oplus \cdots \oplus z_{kn}\colon\ n \in \mathbb{J}\}$ is a Parseval frame for $H_1 \oplus \cdots \oplus H_k$, and we call it a *complete* strongly disjoint k-tuple if $\{z_{1n} \oplus \cdots \oplus z_{kn}\colon\ n \in \mathbb{J}\}$ is an orthonormal basis for $H_1 \oplus \cdots \oplus H_k$. If $\theta_i\colon\ H_i \to l^2(\mathbb{J})$ is the frame transform, $1 \le i \le k$, then strong disjointness of a k-tuple is equivalent to mutual orthogonality of $\{\mathrm{ran}\ \theta_i\colon\ 1 \le i \le k\}$, and complete strong disjointness is equivalent to the condition that $\bigoplus_{i=1}^{k} \mathrm{ran}\ \theta_i = l^2(\mathbb{J})$.

There is a particularly simple intrinsic (i.e. non-geometric) characterization of strong disjointness which is potentially useful in applications: Let $\{x_n\}_{n\in\mathbb{J}}$ and $\{y_n\}_{n\in\mathbb{J}}$ be Parseval frames for Hilbert spaces H and K, respectively. Then $\{x_n\}$ and $\{y_n\}$ are strongly disjoint if and only if one of the equations

$$\sum_{n\in\mathbb{J}}\langle x, x_n\rangle y_n = 0 \quad \text{for all } x \in H \tag{37}$$

$$\text{or} \quad \sum_{n\in\mathbb{J}}\langle y, y_n\rangle x_n = 0 \quad \text{for all } y \in K$$

holds. Moreover, if one holds the other holds also.

3.4. *Super-frames, super-wavelets, and multiplexing*

Suppose that $\{x_n\}_{n\in\mathbb{J}}$ and $\{y_n\}_{n\in\mathbb{J}}$ are strongly disjoint Parseval frames for Hilbert spaces H and K, respectively. Then given any pair of vectors $x \in H$, $y \in K$, we have that

$$x = \sum_n \langle x, x_n\rangle x_n, \qquad y = \sum_n \langle y, y_n\rangle y_n.$$

If we let $a_n = \langle x, x_n\rangle$ and $b_n = \langle y, y_n\rangle$, and then let $c_n = a_n + b_n$, we have

$$\sum_n a_n y_n = 0, \qquad \sum_n b_n x_n = 0,$$

by (37) and therefore we have

$$x = \sum_n c_n x_n, \qquad y = \sum_n c_n y_n. \tag{38}$$

This says that, by using one set of data $\{c_n\}$, we can recover two vectors x and y (they may even lie in different Hilbert spaces) by applying the respective inverse transforms (synthesis operators) corresponding to the two frame $\{x_n\}$ and $\{y_n\}$. The above argument obviously extends to the k-tuple case: If $\{f_{in}\colon\ n \in \mathbb{J}\}$, $i = 1, \ldots, k$, is a strongly disjoint k-tuple of Parseval frames for Hilbert spaces $H_1, \ldots, H_k$, and if $(x_1, \ldots, x_k)$ is an arbitrary k-tuple of vectors with $x_i \in H_i$, $1 \le i \le k$, then (38) generalizes to

$$x_i = \sum_{n\in\mathbb{J}} \langle x_i, f_{in}\rangle f_{in}$$

for each $1 \le i \le k$. So if we define a single "master" sequence of complex numbers $\{c_n\colon\ n \in \mathbb{J}\}$ by

$$c_n = \sum_{i=1}^{k} \langle x_i, f_{in}\rangle,$$

then the strong disjointness implies that for each *individual* i we have

$$x_i = \sum_{n \in \mathbb{J}} c_n f_{in}.$$

This simple observation might be useful in applications to data compression.

In [33] we called such an n-tuple of strongly disjoint (or simply just disjoint) frames a *super-frame*, because it (or rather its inner direct sum) is a frame for the *superspace* which is the direct sum of the individual Hilbert spaces for the frames. In connection with wavelet systems this observation lead us to the notion of *superwavelet*, which is a particular type of vector-valued wavelet. In operator-theoretic terms this is just a restatement of the fact outlined above that a strongly disjoint k-tuple of Parseval frames have frame-transforms which are isometries into the same space $l^2(\mathbb{J})$ which have mutually orthogonal ranges.

The notion of superframes and superwavelets, and many of their properties, were also discovered and investigated by Radu Balan [3] in his Ph.D. thesis, in work that was completely independent from ours.

3.5. *Frame vectors for unitary systems*

Let $\mathcal{U}$ be a unitary system on a Hilbert space H. Suppose $\mathcal{W}(\mathcal{U})$ is nonempty, and fix $\psi \in \mathcal{W}(\mathcal{U})$. Recall from Section 1 that if η is an arbitrary vector in H, then $\eta \in \mathcal{W}(\mathcal{U})$ if and only if there is a unitary V (which is unique if it exists) in the local commutant $\mathcal{C}_\psi(\mathcal{U})$ such that $V\psi = \eta$. The following proposition shows that this idea generalizes to the theory of frames. Analogously to the notion of a wandering vector and a complete wandering vector, a vector $x \in H$ is called a *Parseval frame vector* (resp. *frame vector* with bounds a and b) for a unitary system $\mathcal{U}$ if $\mathcal{U}x$ forms a tight frame (resp. frame with bounds a and b) for $\overline{span}(\mathcal{U}x)$. It is called a *complete Parseval frame vector* (resp. *complete frame vector* with bounds a and b) when $\mathcal{U}x$ is a Parseval frame (resp. frame with bounds a and b) for H.

Proposition 15: *Suppose that ψ is a complete wandering vector for a unitary system $\mathcal{U}$. Then*

(i) *a vector η is a Parseval frame vector for $\mathcal{U}$ if and only if there is a (unique) partial isometry $A \in C_\psi(\mathcal{U})$ such that $A\psi = \eta$.*
(ii) *a vector η is a complete Parseval frame vector for $\mathcal{U}$ if and only if there is a (unique) co-isometry $A \in C_\psi(\mathcal{U})$ such that $A\psi = \eta$.*

The above result does not tell the whole story. The reason is that many unitary systems do not have wandering vectors but do have frame vectors. For instance, this is the case in Example 14, where the unitary system is the group of multiplication operators $\mathcal{U} = \{M_{e^{ins}}\colon\ n \in \mathbb{Z}\}$ acting on $L^2(E)$. In the case of a unitary system such as the wavelet system $\mathcal{U}_{D,T}$ there exist *both* complete wandering vectors *and* nontrivial Parseval frame vectors, so the theory seems richer (however less tractable) and Proposition 15 is very relevant.

Much of Example 14 generalizes to the case of an arbitrary countable unitary group. There is a corresponding (geometric) dilation result.

Proposition 16: *Suppose that $\mathcal{U}$ is a unitary group such that $\mathcal{W}(\mathcal{U})$ is non-empty. Then every complete Parseval frame vector must be a complete wandering vector.*

Theorem 17: *Suppose that $\mathcal{U}$ is a unitary group on H and η is a complete Parseval frame vector for $\mathcal{U}$. Then there exists a Hilbert space $K \supseteq H$ and a unitary group $\mathcal{G}$ on K such that $\mathcal{G}$ has complete wandering vectors, H is an invariant subspace of $\mathcal{G}$ such that $\mathcal{G}|_H = \mathcal{U}$, and the map $g \to g|_H$ is a group isomorphism from $\mathcal{G}$ onto $\mathcal{U}$.*

The following is not hard, but it is very useful.

Proposition 18: *Suppose that $\mathcal{U}$ is a unitary group which has a complete Parseval frame vector. Then the von Neumann algebra $w^*(\mathcal{U})$ generated by $\mathcal{U}$ is finite.*

3.6. *An operator model*

The following is a corollary of Theorem 17. It shows that Example 14 can be viewed as a model for certain operators.

Corollary 19: *Let $T \in B(H)$ be a unitary operator and let $\eta \in H$ be a vector such that $\{T^n\eta\colon\ n \in \mathbb{Z}\}$ is a Parseval frame for H. Then there is a unique (modulo a null set) measurable set $E \subset \mathbb{T}$ such that $\{T^n\eta\colon\ n \in \mathbb{Z}\}$ and $\{e^{ins}|_E\colon\ n \in \mathbb{Z}\}$ are unitarily equivalent frames.*

3.7. *Group representations*

These concepts generalize. For a unitary system $\mathcal{U}$ on a Hilbert space H, a closed subspace M of H is called a *complete wandering subspace* for $\mathcal{U}$ if span $\{UM\colon\ U \in \mathcal{U}\}$ is dense in H, and $UM \perp VM$ with $U \neq V$. Let

$\{e_i\colon\ i \in I\}$ be an orthonormal basis for M. Then M is a complete wandering subspace for $\mathcal{U}$ if and only if $\{Ue_i\colon\ U \in \mathcal{U}, i \in I\}$ is an orthonormal basis for H. We call $\{e_i\}$ a *complete multi-wandering vector.* Analogously, an n-tuple $(\eta_1, \ldots, \eta_n)$ of non-zero vectors (here n can be ∞) is called *complete Parseval multi-frame vector* for $\mathcal{U}$ if $\{U\eta_i\colon\ U \in \mathcal{U}, i = 1, \ldots, n\}$ forms a complete Parseval frame for H. Let G be a group and let λ be the left regular representation of G on $l^2(G)$. Then $\{\lambda_g \times I_n\colon\ g \in G\}$ has a complete multi-wandering vector $(f_1, \ldots, f_n)$, where $f_1 = (x_e, 0, \ldots, 0), \ldots, f_n = (0, 0, \ldots, x_e)$. Let P be any projection in the commutant of $(\lambda \otimes I_n)(\mathcal{G})$. Then $(Pf_1, \ldots, Pf_n)$ is a complete Parseval multi-frame vector for the subrepresentation $(\lambda \otimes I_n)|_P$. It turns out that every representation with a complete Parseval multi-frame vector arises in this way. Item (i) of the following theorem is elementary and was mentioned earlier; it is included for completeness.

Theorem 20: *Let G be a countable group and let π be a representation of G on a Hilbert space H. Let λ denote the left regular representation of G on $l^2(G)$. Then*

(i) *if $\pi(G)$ has a complete wandering vector then π is unitarily equivalent to λ,*
(ii) *if $\pi(G)$ has a complete Parseval frame vector then π is unitarily equivalent to a subrepresentation of λ,*
(iii) *if $\pi(G)$ has a complete Parseval multi-frame vector $\{\psi_1, \psi_2, \ldots, \psi_n\}$, for some $1 \leq n < \infty$, then π is unitarily equivalent to a subrepresentation of $\lambda \otimes I_n$.*

4. Decompositions of Operators and Operator-Valued Frames

The material we present here is contained in two recent papers. The first [16] was authored by a [VIGRE/REU] team consisting of K. Dykema, D. Freeman, K. Kornelson, D. Larson, M. Ordower, and E. Weber, with the title *Ellipsoidal Tight Frames.* This article started as an undergraduate research project at Texas A&M in the summer of 2002, in which Dan Freeman was the student and the other five were faculty mentors. Freeman is now a graduate student at Texas A&M. The project began as a solution of a finite dimensional frame research problem, but developed into a rather technically deep theory concerning a class of frames on an infinite dimensional Hilbert space. The second paper [42], entitled *Rank-one decomposition of operators and construction of frames*, is a joint article by K. Kornelson and D. Larson.

4.1. *Ellipsoidal frames*

We will use the term *spherical frame* (or *uniform frame*) for a frame sequence which is *uniform* in the sense that all its vectors have the same norm. Spherical frames which are tight have been the focus of several articles by different researchers. Since frame theory is essentially geometric in nature, from a purely mathematical point of view it is natural to ask: Which other surfaces in a finite or infinite dimensional Hilbert space contain tight frames? (These problems can make darn good REU projects, in particular.) In the first article we considered ellipsoidal surfaces.

By an *ellipsoidal surface* we mean the image of the unit sphere S_1 in the underlying Hilbert space H under a bounded invertible operator A in $B(H)$, the set of all bounded linear operators on H. Let E_A denote the ellipsoidal surface $E_A := AS_1$. A frame contained in E_A is called an *ellipsoidal frame*, and if it is tight it is called an ellipsoidal tight frame (ETF) for that surface. We say that a frame bound K is *attainable* for E_A if there is an ETF for E_A with frame bound K.

Given an ellipsoidal surface $E := E_A$, we can assume $E = E_T$ where T is a positive invertible operator. Indeed, given an invertible operator A, let $A^* = U|A^*|$ be the polar decomposition, where $|A^*| = (AA^*)^{1/2}$. Then $A = |A^*|U^*$. By taking $T = |A^*|$, we see that $TS_1 = AS_1$. Moreover, it is easily seen that the positive operator T for which $E = E_T$ is unique.

The starting point for the work in the first paper was the following Proposition. For his REU project Freeman found an elementary calculus proof of this for the real case. Others have also independently found this result, including V. Paulsen, and P. Casazza and M. Leon.

Proposition 21: *Let E_A be an ellipsoidal surface on a finite dimensional real or complex Hilbert space H of dimension n. Then for any integer $k \geq n$, E_A contains a tight frame of length k, and every ETF on E_A of length k has frame bound $K = k\left[trace(T^{-2})\right]^{-1}$.*

We use the following standard definition: For an operator $B \in H$, the *essential norm* of B is:

$$\|B\|_{ess} := \inf\{\|B - K\| \ : \ K \text{ is a compact operator in } B(H)\}$$

Our main frame theorem from the first paper is:

Theorem 22: *Let E_A be an ellipsoidal surface in an infinite dimensional real or complex Hilbert space. Then for any constant $K > \|T^{-2}\|_{ess}^{-1}$, E_T contains a tight frame with frame bound K.*

So, for fixed A, in finite dimensions the set of attainable ETF frame bounds is finite, whereas in infinite dimensions it is a continuum.

Problem. If the essential norm of A is replaced with the norm of A in the above theorem, or if the inequality is replaced with equality, then except for some special cases, and trivial cases, no theorems of any degree of generality are known concerning the set of attainable frame bounds for ETF's on E_A. It would be interesting to have a general analysis of the case where $A - I$ is compact. In this case, one would want to know necessary and sufficient conditions for existence of a tight frame on E_A with frame bound 1. In the special case $A = I$ then, of course, any orthonormal basis will do, and these are the only tight frames on E_A in this case. What happens in general when $\|A\|_{ess} = 1$ and A is a small perturbation of I?

We use elementary tensor notation for a rank-one operator on H. Given $u, v, x \in H$, the operator $u \otimes v$ is defined by $(u \otimes v)x = \langle x, v\rangle u$ for $x \in H$. The operator $u \otimes u$ is a projection if and only if $\|u\| = 1$.

Let $\{x_j\}_j$ be a frame for H. The standard frame operator is defined by: $Sw = \sum_j \langle w, x_j\rangle x_j = \sum_j (x_j \otimes x_j)\, w$. Thus $S = \sum_j x_j \otimes x_j$, where this series of positive rank-1 operators converges in the strong operator topology (i.e. the topology of pointwise convergence). In the special case where each x_j is a unit vector, S is the sum of the rank-1 projections $P_j = x_j \otimes x_j$.

For A a positive operator, we say that A has a *projection decomposition* if A can be expressed as the sum of a finite or infinite sequence of (not necessarily mutually orthogonal) self-adjoint projections, with convergence in the strong operator topology.

If x_j is a frame of unit vectors, then $S = \sum_j x_j \otimes x_j$ is a projection decomposition of the frame operator. This argument is trivially reversible, so a positive invertible operator S is the frame operator for a frame of unit vectors if and only if it admits a projection decomposition $S = \sum_j P_j$. If the projections in the decomposition are not of rank one, each projection can be further decomposed (orthogonally) into rank-1 projections, as needed, expressing $S = \sum_n x_n \otimes x_n$, and then the sequence $\{x_n\}$ is a frame of unit vectors with frame operator S.

In order to prove Theorem 22, we first proved Theorem 23 (below), using purely operator-theoretic techniques.

Theorem 23: *Let A be a positive operator in $B(H)$ for H a real or complex*

Hilbert space with infinite dimension, and suppose $\|A\|_{ess} > 1$. *Then* A *has a projection decomposition.*

Suppose, then, that $\{x_n\}$ is a frame of unit vectors with frame operator S. If we let $y_j = S^{-\frac{1}{2}}x_j$, then $\{y_j\}_j$ is a *Parseval* frame. So $\{y_j\}_j$ is an ellipsoidal tight frame for the ellipsoidal surface $E_{S^{-\frac{1}{2}}} = S^{-\frac{1}{2}}S_1$. This argument is reversible: Given a positive invertible operator T, let $S = T^{-2}$. Scale T if necessary so that $\|S\|_{ess} > 1$. Let $S = \sum_j x_j \otimes x_j$ be a projection decomposition of S. Then $\{Tx_j\}$ is an ETF for the ellipsoidal surface TS_1. Consideration of frame bounds and scale factors then yields Theorem 22.

Most of our second paper concerned *weighted* projection decompositions of positive operators, and resultant theorems concerning frames. If T is a positive operator, and if $\{c_n\}$ is a sequence of positive scalars, then a weighted projection decomposition of T with weights $\{c_n\}$ is a decomposition $T = \sum_j P_j$ where the P_j are projections, and the series converges strongly. We have since adopted the term *targeted* to refer to such a decomposition, and generalizations thereof. By a *targeted decomposition* of T we mean any strongly convergent decomposition $T = \sum_n T_n$ where the T_n is a sequence of *simpler* positive operators with special prescribed properties. So a weighted decomposition is a targeted decomposition for which the scalar weights are the prescribed properties. And, of course, a projection decomposition is a special case of targeted decomposition.

After a sequence of Lemmas, building up from finite dimensions and employing spectral theory for operators, we arrived at the following theorem. We will not discuss the details here because of limited space. It is the *weighted* analogue of Theorem 23.

Theorem 24: *Let* B *be a positive operator in* $B(H)$ *for* H *with* $\|B\|_{ess} > 1$. *Let* $\{c_i\}_{i=1}^{\infty}$ *be any sequence of numbers with* $0 < c_i \leq 1$ *such that* $\sum_i c_i = \infty$. *Then there exists a sequence of rank-one projections* $\{P_i\}_{i=1}^{\infty}$ *such that* $B = \sum_{i=1}^{\infty} c_i P_i$.

4.2. *A problem in operator theory*

We will discuss a problem in operator theory that was motivated by a problem in the theory of Modulation Spaces. We tried to obtain an actual "reformulation" of the modulation space problem in terms of operator theory, and it is well possible that such a reformulation can be found. At the least we (Chris Heil and myself) found the following operator theory problem, whose solution could conceivably impact mathematics beyond operator theory. I find it rather fascinating. I need to note that we subsequently showed (in an unpublished jointly-written expository article) that the actual *modulation space* connection requires a modified and more sophisticated version of the problem we present below. I still feel, that the problem I will present here has some independent interest, and may serve as a "first step" in developing a theory that might have some usefulness. Thus, I hope that the reader will find it interesting.

Let H be an infinite dimensional separable Hilbert space. As usual, denote the Hilbert space norm on H by $\|\cdot\|$. If x and y are vectors in H, then $x \otimes y$ will denote the operator of rank one defined by $(x \otimes y)z = \langle z, y \rangle x$. The operator norm of $x \otimes y$ is then just the product of $\|x\|$ and $\|y\|$.

Fix an orthonormal basis $\{e_n\}_n$ for H. For each vector v in H, define

$$|||v||| = \sum_n |\langle v, e_n \rangle|$$

This may be $+\infty$.

Let L be the set of all vectors v in H for which $|||v|||$ is finite. Then L is a dense linear subspace of H, and is a Banach space in the "triple norm". It is of course isomorphic to ℓ^1.

Let T be any positive trace-class operator in $B(H)$.

The usual eigenvector decomposition for T expresses T as a series converging in the strong operator topology of operators $h_n \otimes h_n$, where $\{h_n\}$ is an orthogonal sequence of eigenvectors of T. That is,

$$T = \sum_n h_n \otimes h_n$$

In this representation the eigenvalue corresponding to the eigenvector h_n is the square of the norm: $\|h_n\|^2$. The trace of T is then

$$\sum_n \|h_n\|^2$$

and since T is positive this is also the trace-class norm of T.

Let us say that T is of *Type A* with respect to the orthonormal basis $\{e_n\}$ if, for the eigenvectors $\{h_n\}$ as above, we have that $\sum_n |||h_n|||^2$ is finite. [Note that this is just the (somewhat unusual) formula displayed above for the trace of T with the triple norm used in place of the usual Hilbert space norm of the vectors $\{h_n\}$.]

And let us say that T is of *Type B* with respect to the orthonormal basis $\{e_n\}$ if there is *some* sequence of vectors $\{v_n\}$ in H with $\sum_n |||v_n|||^2$ *finite* such that

$$T = \sum_n v_n \otimes h_n$$

where the convergence of this series is in the strong operator topology.

Problem: If T is of *Type B* with respect to an orthonormal basis $\{e_n\}$, then must it be of *Type A* with respect to $\{e_n\}$?

Note: If the answer to this problem is negative (as I suspect it is), then the following subproblem would be an interesting one.

Subproblem: Let $\{e_n\}$ be an orthonormal basis for H. Find a characterization of all positive trace class operators T that are of *Type B* with respect to $\{e_n\}$. In particular, is every positive trace class operator T of *Type B* with respect to $\{e_n\}$? My feeling is *no.* (See the next example.)

Example 25: Let x be any vector in H that is not in L, and let $T = x \otimes x$. Then T is trace class, in fact has rank one, but clearly T is *not* of Type A. Can such a T be of Type B? (I don't think it is necessarily of Type B for all such T, however.)

References

1. A. Aldroubi, D. R. Larson, W.-S. Tang, and E. Weber, *Geometric aspects of frame representations of abelian groups,* Trans. Amer. Math. Soc. **356** (2004), 4767–4786.
2. E. A. Azoff, E. J. Ionascu, D. R. Larson, and C. M. Pearcy, *Direct paths of wavelets,* Houston J. Math. **29** (2003), no. 3, 737–756.
3. R. Balan, *A study of Weyl-Heisenberg and wavelet frames,* Ph.D. thesis, Princeton University, 1998.
4. L. Baggett, H. Medina, and K. Merrill, *Generalized multi-resolution analyses and a construction procedure for all wavelet sets in R^n*, J. Fourier Anal. Appl. **5** (1999).
5. J. Benedetto and M. Fickus, Finite normalized tight frames, Adv. Comput. Math., 18 (2003), 357-385.
6. J. J. Benedetto and M. Leon, *The construction of single wavelets in D-dimensions,* J. Geom. Anal. **11** (2001), no. 1, 1–15.

7. P. Casazza, D. Han and D. Larson, *Frames for Banach spaces,* Contemp. Math., **247**, Amer. Math. Soc., Providence, RI, 1999.
8. O. Christensen, *An introduction to frames and Riesz bases*, Applied and Numerical Harmonic Analysis, Birkhäser, Boston, MA, 2003.
9. X. Dai, *Norm principal bimodules of nest algebras,* J. Functional Analysis, **90** (1990), 369–390.
10. X. Dai, Y. Diao, Q. Gu and D. Han, *Wavelets with frame multiresolution analysis,* J. Fourier Analysis and Applications, **9** (2003), 39-48.
11. X. Dai and D. Larson, *Wandering vectors for unitary systems and orthogonal wavelets,* Mem. Amer. Math. Soc. **134** (1998).
12. X. Dai, D. Larson and D. Speegle, *Wavelet sets in R^n,* J. Fourier Anal. Appl. **3** (1997), no. 4, 451–456.
13. X. Dai, D. Larson and D. Speegle, *Wavelet sets in R^n - II*, Contemp. Math, **216** (1998), 15-40.
14. I. Daubechies, *Ten Lectures on Wavelets*, Society for Industrial and Applied Mathematics (SIAM), Philadelphia, PA, 1992.
15. D. E. Dutkay, *The local trace function for super-wavelets*, Wavelets, Frames, and Operator Theory, Contemp. Math., vol. **345**, (2004), pp. 115–136.
16. K. Dykema, D. Freeman, K. Kornelson, D. Larson, M. Ordower, and E. Weber, *Ellipsoidal tight frames and projection decompositions of operators*, Illinois J. Math. **48** (2004), no. 2, 477–489.
17. X. Fang and X. Wang, Construction of minimally-supported frequencies wavelets, J. Fourier Anal. Appl. 2 (1996), 315-327.
18. H. Feichtinger, *Atomic characterization of modulation spaces through Gabor type representations,* Rocky Mountain J. Math., **19** (1989), 113–126.
19. M. Frank and D. R. Larson, *Frames in Hilbert C*-modules and C*- algebras,* J. Operator Theory **48** (2002), no. 2, 273–314.
20. M. Frank and D. R. Larson, *A module frame concept for Hilbert C*-modules,* Contemporary Mathematics, **247** (1999), 207-234.
21. M. Frank and D. R. Larson, *Frames for Hilbert C* Modules,* SPIE Proceedings Vol. 4119, Wavelet Applications in Signal And Image Processing VIII, (2000), 325-336.
22. M. Frank, V. I. Paulsen and T. R. Tiballi , *Symmetric approximation of frame,* Trans. Amer. Math. Soc., **354** (2002), 777–793.
23. J.-P. Gabardo, D. Han, and D. Larson, *Gabor frames and operator algebras*, Wavelet Applications in Signal and Image Processing, Proc. SPIE, vol. 4119, 2000, pp. 337–345.
24. J.-P. Gabardo and D. Han, *Subspace Weyl-Heisenberg frames*, J. Fourier Analysis and Appl., **7**(2001), 419-433.
25. T. N. T. Goodman, S. L. Lee and W. S. Tang, *Wavelets in wandering subspaces,* Trans. Amer. Math. Soc., **338** (1993), 639-654.
26. Q. Gu, *On interpolation families of wavelet sets,* Proc. Amer. Math. Soc., **128** (2000), 2973–2979.
27. Q. Gu and D. Han, *On multiresolution analysis wavelets in R^n*, J. Fourier Analysis and Applications, **6**(2000), 437-448.
28. Q. Gu and D. Han, *Phases for dyadic orthonormal wavelets,* J. of Mathe-

matical Physics, **43** (2002), no. 5, 2690–2706.

29. Q. Gu and D. Han, *Functional Gabor frame multipliers*, J. Geometric Analysis, **13** (2003), 467–478.
30. D. Han, *Wandering vectors for irrational rotation unitary systems,* Trans. Amer. Math. Soc., **350** (1998), 309-320.
31. D. Han, *Tight frame approximation for multi-frames and super-frames,* J. Approx. Theory, **129** (2004), 78–93
32. D. Han, J.-P. Gabardo, and D. R. Larson, *Gabor frames and operator algebras*, Wavelet Applications in Signal and Image Processing, Proc. SPIE, 4119 (2000), 337-345.
33. D. Han and D. R. Larson, *Frames, bases and group representations*, Memoirs American Math. Society, **697**, (2000).
34. D. Han and D. R. Larson, *Wandering vector multipliers for unitary groups,* Trans. Amer. Math. Soc., **353** (2001), 3347–3370.
35. D. Han and Y. Wang, *The existence of Gabor bases and frames*, Contemp. Math., **345** (2004), 183–192.
36. C. Heil, P. E. T. Jorgensen, and D. R. Larson (eds.), *Wavelets, Frames and Operator Theory,* Contemp. Math., vol. 345, American Mathematical Society, Providence, RI, 2004, Papers from the Focused Research Group Workshop held at the University of Maryland, College Park, MD, January 15–21, 2003.
37. E. Hernandez, X. Wang and G. Weiss, Smoothing minimally supported frequency (MSF) wavelets: Part I., J. Fourier Anal. Appl. 2 (1996), 329-340.
38. E. Hernandez and G. Weiss, *A First Course on Wavelets*, CRC Press, Inc., 1996.
39. E. Ionascu, D. Larson and C. Pearcy, *On the unitary systems affiliated with orthonormal wavelet theory in n-dimensions,* J. Funct. Anal. **157** (1998), no. 2, 413–431.
40. E. Ionascu, D. Larson and C. Pearcy, *On wavelet sets,* J. Fourier Analysis and Applications, **4** (1998), 711–721.
41. R. Kadison and J. Ringrose, *Fundamentals of the Theory of Operator Algebras, Vol. I and II* , Academic Press, Inc. 1983 and 1985.
42. K. Kornelson and D. Larson, *Rank-one decomposition of operators and construction of frames*, Wavelets, Frames, and Operator Theory, Contemp. Math, vol. 345, Amer. Math. Soc., 2004, pp. 203–214.
43. D. R. Larson, *Von Neumann algebras and wavelets. Operator algebras and applications* (Samos, 1996), 267–312, NATO Adv. Sci. Inst. Ser. C Math. Phys. Sci., 495, Kluwer Acad. Publ., Dordrecht, 1997.
44. D. R. Larson, *Frames and wavelets from an operator-theoretic point of view,* Operator algebras and operator theory (Shanghai, 1997), 201–218, Contemp. Math., 228, Amer. Math. Soc., Providence, RI, 1998.
45. D. R. Larson, E. Schulz, D. Speegle and K. Taylor, *Explicit cross sections of singly generated group actions*, to appear.
46. D. R. Larson, W.-S. Tang, and E. Weber, *Riesz wavelets and multiresolution structures*, SPIE Proc. Vol. 4478, Wavelet Applications in Signal and Image

Processing IX (2001), 254-262.

47. D. R. Larson, W. S. Tang, and E. Weber, *Multiwavelets associated with countable groups of unitary operators in Hilbert spaces*, Int. J. Pure Appl. Math. **6** (2003), no. 2, 123–144.
48. G. Ólafsson and D. Speegle, *Wavelets, wavelet sets, and linear actions on* $\mathbb{R}^n$, Wavelets, frames and operator theory, Contemp. Math., vol. 345, Amer. Math. Soc., Providence, RI, 2004, pp. 253–281.
49. D. Speegle, *On the existence of wavelets for non-expansive dilation matrices,* Collect. Math., **54** (2003), 163–179.
50. D. Speegle, *The s-elementary wavelets are path-connected,* Proc. Amer. Math. Soc., **132** (2004), 2567–2575.
51. P. Wood, *Wavelets and Hilbert modules,* to appear in the Journal of Fourier Analysis and Applications (2004).
52. Wutam Consortium, *Basic properties of wavelets* J. Fourier Analysis and Applications, **4** (1998), 575-594.